"Are Christians using technology to transform the world or is technology transforming Christians in unhealthy ways? Especially since the era of Franklin and Jefferson, when inventing things and technological ways of organizing things became a way of life, Christians have needed to be alert to such questions. Tony Reinke's reflections on the smartphone offer helpful advice as to how people today need to be vigilant regarding the impact of their favorite new technologies."

> **George M. Marsden,** Francis A. McAnaney Professor of History Emeritus, University of Notre Dame

"*12 Ways Your Phone Is Changing You* is an incredibly convicting and profoundly insightful read. Smartphones have become a part of our lives, but Tony explores the devastation to the human mind and soul due to devotion to technology. He calls us to examine not merely the use of our smartphones but the motives that inspire it. This is a necessary book for our generation, to remind us that our phone habits will either amplify or get in the way of our most important longing of all: the soul-satisfying glory of our Savior."

> **Jackie Hill Perry,** poet; hip-hop artist

"In contrast to the television that dominates the modern living room, the smartphone is typically far less conspicuous in its presence. Perhaps on account of this subtle unobtrusiveness, surprisingly few have devoted sustained reflection to the effect this now ubiquitous technology is having on our lives. In this book, Tony Reinke plucks these devices from the penumbra of our critical awareness and subjects them to the searching light of Christian wisdom. The result is an often sobering assessment of the effect they are having on our lives, accompanied by much prudent and practical counsel for mastering them. This is a timely and thoughtful treatment of a profoundly important issue, a book that should be prescribed to every Christian smartphone owner for the sake of our spiritual health."

> **Alastair Roberts,** theologian, blogger

"Tony Reinke's *12 Ways Your Phone Is Changing You* is one of the most important little books a twenty-first-century Christian could read. Highly recommended."

> **Bruce Riley Ashford,** Provost and Professor of Theology and Culture, Southeastern Baptist Theological Seminary

"For many, the phone is an object of increasing anxiety, exhaustion, and dependency. The wise Tony Reinke leads us practically to find freedom from the phone without requiring us to huddle away in a monastery somewhere in the middle of Montana. If you want to know how to steward your technology and your life for Christ and his kingdom, read this."

Russell Moore, President, Ethics & Religious Liberty Commission of the Southern Baptist Convention

"If you feel uneasy about your constant relationship with your phone (and even if you don't, but wonder if you should), you will find Tony Reinke to be a reliable guide for how we should assess the impact of our phones on ourselves and our relationships. A marvelous book that tackles a massive subject in clear and compelling language!"

Trevin Wax, Managing Editor, The Gospel Project; author, *Counterfeit Gospels* and *Holy Subversion*

"Two things strike me about this book. First, Reinke writes with great humility, including himself in the narrative to help us see him not only as a teacher but also as a fellow struggler. Second, this is not a guilt-ridden slog through what not to do. Tony keeps pulling us up into the glories of Christ and even helps us to dream of new ways to glorify God through our digital technologies. Helpful, hopeful, humbling, and inspiring, *12 Ways Your Phone Is Changing You* is a book for this age and wisdom for generations to follow."

Trillia Newbell, author, *Enjoy, Fear and Faith,* and *United*

"Image is everything, and for a woman who has built her identity on the sands of how she's embraced online, the eventual letdown will come like a crash. But there's a better way forward, a way to use our phones in selfless service, to glorify God in our connectivity, and to image Christ by our phone behaviors. For this, we must evaluate our glowing screens and train our discernment to see the difference between the sight-driven habits of our age and the Scripture-lit pathway of faith. Every chapter of this book is like the right kind of push notification in our lives. Stop, read, process, and apply with care."

Gloria Furman, author, *Missional Motherhood*

"As a teenager and a smartphone user, I needed this book. Tony Reinke is compelling and convicting, yet continually meets us with grace. My generation needs this book, because we need to get technology right. If we don't, the cost is great. *12 Ways Your Phone Is Changing You* should be a must-read for every smartphone user, especially for us younger ones."

Jaquelle Crowe, author, *This Changes Everything*

"It took more than a generation for the quaint 'horseless carriage,' with all its magic and horror, to become the ordinary, unexamined 'car.' But the device we once called a 'smartphone' has reached its status as 'phone'—a common, everyday inevitability—with such breathtaking speed that it has left us little time for reflection on the true power it has in our lives. Tony offers us a distinctly Christian take on the little wonders in our pockets, seeing their goodness, beauty, and power, but also applying godly wisdom and well-researched cautions to help readers use their phones without being used by their phones."

John Dyer, author, *From the Garden to the City: The Redeeming and Corrupting Power of Technology*

"Experience practical theology at its finest as Tony applies a thorough understanding of the Scriptures to a thorough understanding of our culture, resulting in a beautifully written and balanced guide to the dangers and opportunities in the palms of our hands. Yes, our phones have changed us for the worse, but this book will change us and our phone use for the better."

David Murray, pastor; author; Professor of Old Testament and Practical Theology, Puritan Reformed Theological Seminary

"The more widespread and influential something is, the more Christians should think carefully about it. In this wisdom-filled book, Tony Reinke helps us do just that with the smartphone. Without descending into technophobia or paranoia, he shows the various ways in which phones are changing our lives, highlighting both the problems with this and the solutions to it. A timely and thoughtful book."

Andrew Wilson, author; speaker; Teaching Pastor, King's Church London

"Rarely is a book as practically impactful as it is theologically rich. In an age in which daily we are drawn into a digital vortex, Tony Reinke warns of the implications and challenges us to examine whether our phones have displaced our spiritual priorities in Christ. With unflinching honesty, Reinke shares his own technological struggles, and in so doing, moves us to a posture of reflection, prayer, and even repentance. Thoroughly engaging and immediately applicable, *12 Ways Your Phone Is Changing You* is a must-read for our time."

Kim Cash Tate, author, *Cling: Choosing a Lifestyle of Intimacy with God*

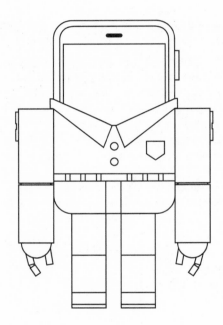

12 WAYS YOUR PHONE IS CHANGING YOU

Tony Reinke

CROSSWAY®

WHEATON, ILLINOIS

Library of Congress Cataloging-in-Publication Data

Names: Reinke, Tony, 1977– author.
Title: 12 ways your phone is changing you / Tony Reinke; foreword by John Piper.
Other titles: Twelve ways your phone is changing you
Description: Wheaton: Crossway, 2017. | Includes bibliographical references and index.
Identifiers: LCCN 2016032849 (print) | LCCN 2016036242 (ebook) | ISBN 9781433552434 (tp) | ISBN 9781433552441 (pdf) | ISBN 9781433552458 (mobi) | ISBN 9781433552465 (epub)
Subjects: LCSH: Technology—Religious aspects—Christianity. | Smartphones.
Classification: LCC BR115.T42 R45 2017 (print) | LCC BR115.T42 (ebook) | DDC 261.5/6—dc23
LC record available at https://lccn.loc.gov/2016032849

Crossway is a publishing ministry of Good News Publishers.

B P		25	24	23	22	21	20	19	18
14	13	12	11	10	9	8	7	6	5

To Karalee

"All things are lawful for me,"
but not all things are helpful.
"All things are lawful for me,"
but I will not be dominated by anything. . . .
"All things are lawful,"
but not all things build up.

—Apostle Paul

CONTENTS

FOREWORD

By John Piper

Smartphones are dangerous, like marriage and music and fine cuisine—or anything else that can become an idol. They are also very useful, like guns and razor blades and medicinal cannabis—or lots of other things that can ruin your life. I personally like marriage very much, and use a razor blade every day. So I am with Tony Reinke in his chastened enthusiasm about the ever-changing world of modern technology.

But I could never have written this book. I don't have the patience, and I don't read fast enough or widely enough. Tony has done more research for this book than for anything else he has written. And those other books were *not* thrown together. His commitment to being informed, and being fair, demanded remarkable attentiveness to subtleties and persistent commitment to ever-clearer reedits. Add to this the gift of theological insightfulness, and this book becomes something very few people could have written. I surely couldn't.

But I do have one small advantage in pondering smartphones. I'm seventy years old. This is an advantage for two reasons. One is that I've been an adult during the entire computer revolution—from the beginning. The other is that I can feel the onrush of eternity just over the horizon.

I got my first real job as a teacher in 1974. I was twenty-eight. The first personal computer was introduced in 1975. It was a kit. I don't do kits. I wait. In 1980, I left academia and became a pastor. Virtually no churches used computers in 1980. They were more like expensive toys and fancy calculators.

But things soon began to get serious. IBM produced its first personal computer in 1981, and *Time* magazine called 1982 "The Year of the Computer." Pricing was prohibitive. But I wanted in for one main reason: word processing. Writing. The price was right in 1984, and my journal entry for June 16 reads: "I bought a computer yesterday. IBM PC, 256K of RAM, double disc for $1,995.00." The monitor was extra. The disk operating system (DOS 2.1) was $60.

Twenty-three years later the iPhone was created. Computer and phone were now one. I was on board within a year. Calling. Texting. Keeping up with the news. Playing Scrabble with my wife. And reading my Bible, saving verses, memorizing on the go. For all the abuses and all the devastation of distraction, wasted hours, narcissistic self-promotion, and pornographic degradation, I see the computer and the smartphone as gifts of God—like papyrus and the codex and paper and the printing press and the organs of mass distribution.

If you live long enough, pray earnestly, and keep your focus on the imperishable Word of God, you can be spared the slavery to newness. Over time, you can watch something wonderful happen. You can see overweening fascination give way to sober usage. You can watch a toy become a tool; a craze become a coworker; a sovereign become a servant. To cite Tony's words—and his aim—you can watch the triumph of useful efficiency over meaningless habit.

I wish I could give every young adult the taste of eternity that grows more intense as you enter your eighth decade. A happy consciousness of the reality of death and the afterlife is a wonderful liberator from faddishness and empty-headed screen-tapping. I say "happy consciousness" because, if all you have is fear, your smartphone almost certainly becomes one of the ways you escape the thought of death.

But if you rejoice in the hope of the glory of God because your sins are forgiven through Jesus, then your smartphone becomes a kind of friendly pack mule on the way to heaven. Mules are not kept for their good looks. They just get the job done.

The job is not to impress anybody. The job is to make much of Christ and love people. That is why we were created. So don't waste your life grooming your mule. Make him bear the weight of a thousand works of love. Make him tread the heights with you in the mountains of worship.

If that sounds strange to you, but perhaps attractive, Tony will serve you well in the pages ahead. Where else will you find the iPhone linked to the New Jerusalem? Where else will someone be wise enough to say that "our greatest need in the digital age is to behold the glory of the unseen Christ in the faint blue glow of our pixelated Bibles"? Where else will we hear fitting praise of Bible apps along with the honest confession that "no app can breathe life into my communion with God"? Who else is writing about the smartphone with the conviction that "the Christian imagination is starving to death for solid theological nourishment"? And who else is going to confront the presumed hiddenness of our private sins with the truth: "There is no such thing as anonymity. It is only a matter of time"?

Yes. And the time is short. Don't waste it parading your mule. Make him work. His Maker will be pleased.

PREFACE

This blasted smartphone! Pesk of productivity. Tenfold plague of beeps and buzzing. Soulless gadget with unquenchable power hunger. Conjuror of digital tricks. Surveillance bracelet. Money pit. Inescapable tether to work. Dictator, distractor, foe!

Yet it is also my untiring personal assistant, my irreplaceable travel companion, and my lightning-fast connection to friends and family. VR screen. Gaming device. Ballast for daily life. My intelligent friend, my alert wingman, and my ever-ready collaborator. This blessed smartphone!

My phone is a window into the worthless and the worthy, the artificial and the authentic. Some days I feel as if my phone is a digital vampire, sucking away my time and my life. Other days, I feel like a cybernetic centaur—part human, part digital—as my phone and I blend seamlessly into a complex tandem of rhythms and routines.

IPHONE 1.0

Tech wiz Steve Jobs introduced the iPhone at Macworld Expo on January 9, 2007, as a "giant" 3.5-inch high-res screen requiring no physical keyboard or stylus. Unlike the clunky smartphones to date, he announced: "We're going to use the best pointing device in the world. We're going to use a pointing device that we're all born with— born with ten of them. We're going to use our fingers." From that moment, the magic of multitouch technology would introduce highly accurate fingertip gestures to a pocket device, bringing humans into more intimate proximity to their computing technology than ever before. When Jobs later announced, as an aside, "You can now *touch* your music," the magnitude of the statement was too mystical to grasp in the moment.[1]

Apple officially released the first iPhone on June 29, 2007, and I bought one that fall. I marveled at the technology stuffed inside this glossy handheld phone: a legitimate computer operating system, a newly engineered iPod for my music, a rapid new mechanism to text friends, super-sharp video combined with a new mobile browser to preserve the full look of the web, an accelerometer to sense how I tip and twist and rotate my phone—all on a screen with intuitive tactile controls guided by fingertip taps, swipes, and pinches.

On a road trip a few days after the sacred unboxing, I stood outside a snowy Iowa rest stop, unlocked my new iPhone, and replied to my first rural email. Wirelessly. Effortlessly. I was hooked, and so were millions of others. In ten years, nearly one billion iPhones have been sold.

Apple's mobile phone was followed by Android, and smartphones spread over the globe and over every corner of our lives. We now check our smartphones every 4.3 minutes of our waking lives.[2] Since I got my first iPhone, a smartphone has been within my reach 24/7: to wake me in the morning, to deejay my music library, to entertain me with videos, movies, and live television, to capture my life in digital pictures and video, to allow me to play the latest video game, to guide me down foreign streets, to broadcast my social media, and to reassure me every night that it will wake me again (as long as I feed it electricity). I use my phone to keep our always-changing family schedule in real-time sync. I used my phone to research, edit, and even write sections of this book. I use my phone for just about everything (except phone calls, it seems). And my phone goes with

---

1. Mic Wright, "The Original iPhone Announcement Annotated: Steve Jobs' Genius Meets Genius," The Next Web, thenextweb.com (Sept. 6, 2015).

2. Jacob Weisberg, "We Are Hopelessly Hooked," *The New York Review of Books* (Feb. 25, 2016).

me wherever I go: the bedroom, the office, vacation, and, yes, the bathroom.

The smartphone combined several budding technologies[3] into the most powerful handheld tool of social connection ever invented. With our phones, all of life is immediately capturable and shareable. So I was not surprised when the editors of *Time* named the iPhone the single most influential gadget of all time, saying that it "fundamentally changed our relationship to computing and information—a change likely to have repercussions for decades to come."[4]

Oh, yes, the repercussions. What is the price of all this digital magic? I have since discovered that my omnipresent iPhone is also corroding my life with distractions—something Apple execs unwittingly admitted on the eve of the launch of the Apple Watch, marketed as a newer and less-invasive techno-fix to all the techno-noise brought into our lives by the iPhone.[5]

Unknown to me at the time I was unboxing my first iPhone, Jobs was actively shielding his children from his digital machines.[6]

Should I be shielding myself?

THE BIG QUESTION

The makers and marketers of the smartphone wield great power over us, and I want to know what effect this technology has on my spiritual life. As in every area of the Christian life, I want to learn from the history of the church and from older Christians. My first interview of many in the path of producing this book was a phone

---

3. This book is far too short to retell the riveting history of the smartphone. For that, see Majeed Ahmad, *Smartphone: Mobile Revolution at the Crossroads of Communications, Computing and Consumer Electronics* (North Charleston, SC: CreateSpace, 2011).

4. Lisa Eadicicco et al., "The 50 Most Influential Gadgets of All Time," *Time* magazine (May 3, 2016).

5. David Pierce, "iPhone Killer: The Secret History of the Apple Watch," *Wired* (April 2015).

6. In 2010, just after Apple launched its innovative tablet (the iPad), a reporter asked Jobs, "So, your kids must love the iPad?" He responded: "They haven't used it. We limit how much technology our kids use at home." Nick Bilton, "Steve Jobs Was a Low-Tech Parent," *The New York Times* (Sept. 10, 2014). Later, Apple's vice president of design, Jonathan Ive, admitted to setting "strict rules about screen time" for his ten-year-old twin boys. Ian Parker, "The Shape of Things to Come," *The New Yorker* (March 2, 2015).

call to seventy-five-year-old theologian David Wells (1939–). His most recent book on God's holiness was surprisingly filled with talk about technology (a relevant subtopic now in any conversation).[7]

"It is only since the mid 1990s that the web has been widely used in our society, so we are talking here about two decades," Wells told me. "And so we—all of us—are trying to figure out what is useful to us and what damages us. We can't escape it, and probably none of us *wants* to escape it. We cannot become digital monks." To my surprise, Wells seemed personally familiar with the temptations: "There is no doubt that life is more highly distracted, because we get pings and beeps and text messages. We are, in fact, living with a parallel, virtual universe, a universe that can take all of the time that we have. What happens to us when we are in constant motion—when we are almost addicted to constant visual stimulation? What is this doing to us? That is the big question."[8]

Wells is exactly right—our phones are constant variables, always changing and morphing new behaviors in us. Many years ago, Jacques Ellul (1912–1994) prophetically warned of this danger of the technological age, writing that "unpredictability is one of the general features of technological progress."[9] The unpredictability of the tech age carries with it a certain level of unabated insecurity that pushes us far from an answer to Wells's question. We don't know what our smartphones are doing to us, but we are being changed, that much is clear.

I later emailed seventy-one-year-old Oliver O'Donovan (1945–), an accomplished Christian ethicist in Scotland, to ask him if Christians should feel uneasy about the rise of digital communications technology. "Electronic communications are a question for the younger generation more than for mine," he admitted. "It is they who have really to learn to understand the powers and threats that they embody, partly through trial and error, but also, and very importantly,

---

7. David Wells, *God in the Whirlwind: How the Holy-love of God Reorients Our World* (Wheaton, IL: Crossway, 2014).
8. David Wells, interview with the author via phone (July 9, 2014).
9. Jacques Ellul, *The Technological Bluff* (Grand Rapids, MI: Eerdmans, 1990), 60.

through remembering what was of greatest importance *before* the communications revolution kicked in.

"Nobody has ever had to learn this before," he said of the questions we now face. "Nobody can teach the rising generation how to learn it. It is a massive challenge to conscientious intelligence, handed uniquely to *them*. The danger they face, of course, is that the tools set the agenda. A tool of communication is a tool *for communicating something*." He then echoed the question from Wells: "Media don't just lie around passively, waiting for us to come along and find them useful for some project we have in mind. They tell us what to do and, more significantly, what to *want* to do. There is a current in the stream, and if we don't know how to swim, we shall be carried by it. I see someone doing something and I want to do it, too. Then I forget whatever it was that I thought I wanted to do."

O'Donovan concluded the interview with a striking warning: "This generation has the unique task assigned it of discerning what the new media are *really good for*, and that means, also, what they are *not* good for. If they fluff it, generations after them will pay the price."[10]

MY TENSIONS

I wanted to write this book in conversation with elders in the church, but my questions for Wells and O'Donovan boomeranged a question back at me: How can we who are most familiar with our smartphones do our best to flesh out the consequences?

I also find myself in a tricky place—asking critical questions about how my phone is changing me while also working full time online and trying to leverage my skills and experiences to grab the attention of a virtual audience. As the online world is growing global, and growing mobile, new gospel opportunities are opening, too.

Broadly speaking, the power of the digital age to pool human intelligence and factual data is unprecedented (Wikipedia is only one

---

10. Oliver O'Donovan, interview with the author via email (Feb. 10, 2016).

example of what's to come). Every Christian is now given unmatched opportunities for online ministry. Our prominent preachers today can reach hundreds of thousands of people through social media. Even the most average Christian can speak to an immediate audience of two hundred or three hundred friends on Facebook, a reach unparalleled in human history.

So I feel the squeeze of this catch-22. I want to become skilled at winning attention online (for Christ), but I also want to ask critical questions about my own phone impulses, habits, and assumptions.

MY INTENTION

This book about phones could easily grow thicker than a phone book, so to keep it short, I must address only the essentials and navigate with care and brevity. While some writers claim our phones are making us cognitively sharper and relationally deeper,[11] others warn that our phones are making us shallow, dumb, and less competent in the real world.[12] Both arguments ring true at times, but "social media are largely what we make of them—escapist or transforming depending on what we expect from them and how we use them."[13] The question of this book is simple: What is the best use of my smartphone in the flourishing of my life?

To that end, my aim is to avoid both extremes: the utopian optimism of the technophiliac and the dystopian pessimism of the technophobe. O'Donovan is exactly right when he says that our temptation is to watch someone doing something and then merely to copy the behavior and lose sight of our personal callings and life goals. In other words, we must ask ourselves: What technologies serve my aims? And what are my goals in the first place? Without

---

11. Clive Thompson, *Smarter Than You Think: How Technology Is Changing Our Minds for the Better* (New York: Penguin, 2013) and Steven Johnson, *Everything Bad Is Good for You: How Today's Popular Culture Is Actually Making Us Smarter* (New York: Riverhead Books, 2006).

12. Nicholas Carr, *The Shallows: What the Internet Is Doing to Our Brains* (New York: W. W. Norton, 2011) and Mark Bauerlein, *The Dumbest Generation: How the Digital Age Stupefies Young Americans and Jeopardizes Our Future (Or, Don't Trust Anyone Under 30)* (New York: TarcherPerigee, 2009).

13. Andy Crouch, *Strong and Weak: Embracing a Life of Love, Risk & True Flourishing* (Downers Grove, IL: InterVarsity Press, 2016), 87.

clear answers here, we can make no progress in thinking through the pros and cons of smartphones *as Christians*.

And yet, if you own a smartphone, you have likely abused it. Such abuse is the target of countless magazine features, books of lament, and powerful videos that reveal just how foolishly our smartphone overuse influences our lives. A moment of guilt can be a powerful motivator, but it won't last. As time wears on and guilt subsides, we revert to old behaviors. This is because our fundamental convictions are too flimsy to sustain new patterns of behavior, and so what seems immediately "right" (turning off our phones) is really nothing more than the product of a moment's worth of shame. What we need are new life disciplines birthed from a new set of life priorities and empowered by our new life freedom in Jesus Christ. So I cannot tell you to put your phone away, to give it up, or to take it up again after a season of burnout. My aim is to explore why you would consider such actions in the first place.

SMALL PRINT

Here are a handful of notes to keep in mind as we begin.

First, this book is written *to me* as much as it is written *by me*. Not only do I need this message, I bear its greatest burden. If the title seems to imply that I'm preaching at you, I'm not. I'm preaching at me. Not many of you should become authors, for we who write books of ethics are held to our words more strictly than anyone.

Second, to keep this book's title short, I have implied that everything in this book is relevant for every individual reader. In truth, I have never been more aware of the variety of smartphone behaviors. We grab our phones as content creators or content consumers, and we focus on timeless content or timely content. Likewise, our smartphone relationships trend in certain directions: as part of virtual communities or as complements to our face-to-face relationships. And those conversations constantly drift toward edification or chitchat (see Figure 1, p. 22). All of us are sliding around these grids constantly, and each trend has its own strengths and pitfalls to address in the pages ahead. But none of us can plot ourselves exactly in the same spot. I mention this at the front of the book as a way to ask for patience when we discuss behaviors that may not immediately apply to you.

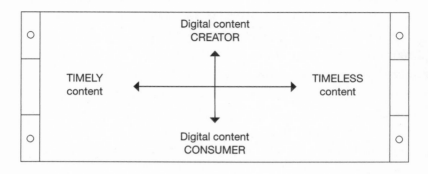

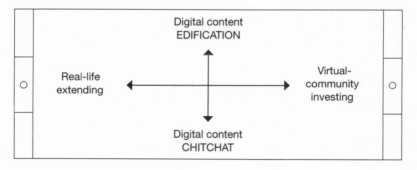

Figure 1. Smartphone behaviors and relationships

Third, this book is not antismartphone; it was written for people who, like me, benefit from the smartphone and use it daily. You will probably hear about this book on your phone in social media, and some of you will read this book on your phones, maybe even quote from it on Facebook—that's not oxymoronic, ironic, or paradoxical; it's the fulfillment of why I wrote it and how I intend to get the message out.

Fourth, this book is not prosmartphone, either. I want this book to be balanced, but balance is not my driving concern. Whether or not I strike the prophone/antiphone balance throughout (or even section by section) is of little concern because I know that, in the end, readers will be split. I concede this point up front in order to speak more directly to my readers who intend to rethink life patterns (and to avoid bloating this book with a million conditions, caveats, and qualifications). I proceed under the assumption that we all need to stop and reflect on our impulsive smartphone habits because, in an age when our eyes and hearts are captured by the latest polished gadget, we need more self-criticism, not less.

Fifth, since you are reading a book titled 12 *Ways Your Phone Is Changing You*, I assume you are likely the type of reader who bravely welcomes such self-critique. I applaud you for it. The old philosopher Seneca was exactly right when he said, "Be harsh with yourself at times."[14] Sometimes. Not always. At certain key moments in life, lean into the bathroom mirror, squint your eyes, and project pessimism at the person you see. We all need healthy critique. But if you are *only* harsh with yourself, let me speak a word of caution. This book fails if, having read it, you only hate yourself more; it succeeds only if you enjoy Christ more. So if you are easily weighed down with conviction and self-doubt, I pray that this book educates and equips you to enjoy freedom in life to taste deeper the infinite joy we have in Christ, leaving mediocre indulgences behind for deeper and more satisfying pleasures ahead.

Sixth, I'll be quoting theologians, philosophers, professors, pastors, popes, perceptive non-Christians, and public atheists—which means that inclusion in this book is not a full endorsement of someone's theology or a wholesale endorsement of the links, apps, books, or mobster movies mentioned ahead.

Finally, as the title suggests, this book centers on diagnostics and worldview more than application. We won't ignore important practices, but the application will be implied generically throughout and addressed specifically at the end.

---

14. Seneca, *Letters from a Stoic: Epistulae Morales ad Lucilium*, trans. Robin Campbell (New York: Penguin, 2015), 67.

CALL FOR HUMILITY

Self-doubt is a hallmark of wise creatures.[15] And self-critical conversations about our personal behaviors require a big dose of humility. Conversations about our smartphones often do not raise new questions; they return us to perennial questions every generation has been forced to ask.

Take Snapchat, the latest phenomenon in "instant expression." In one of my interviews, a theologian suggested to me that it is difficult to let your "yes" be *yes* when your words disappear in a few seconds.[16] But defensive techies immediately negate this claim with a simple fact: while ephemeral words shared on Snapchat disappear in seconds, our vocalized words disappear from the air in *hundredths of a second*. Technology does not make our words more temporary—if anything, it makes them more durable. If we must give an account of every idle word, we are probably the first generation that can truly appreciate the volume of our idle words, since we have published more of them than any group in human history.

So although we can examine our authenticity when we speak through intentionally self-destructing messages (such as Snapchat), our phones do not make our words more transient or empty; they merely raise questions asked in every generation. Only when we acknowledge these questions can we then get back to examining Snapchat.

That is often how conversations on digital media work. So I begin the book by asking for a truce. Can we agree that some of the most important smartphone questions will also apply to nondigital conversations? Just because a struggle we face in our digital lives also relates to nondigital contexts does not mean that the conversation with digital communication is averted—it means that Scripture proves its ongoing relevance in the digital age.

---

15. Prov. 3:5–8; 12:15; 26:12.
16. James 5:12.

WHO AM I?

As you can see, this journey to untangle my relationship with my phone is very personal (i.e., self-critical of *me*), so you need to know who I am from the outset.

I'm "an early adopter"—a nice way of saying "self-professed iPhone addict and techno-junkie." I am also a Christian of nearly two decades who holds the Bible as the ultimate and final author-ity over my life. Educated in business, journalism, and liberal arts, I now work as an investigative reporter of the complex dynamics of the Christian life in tension with the current pressures of cultural conformity. I research and write in concert with many other voices in the church, both living and dead.

Married for nearly two decades, my wife and I have three kids, and we are trying to raise them to be technologically competent and digitally self-controlled.[17] In our home, we currently run one desktop computer, three laptops, three tablets, three smartphones, and one iPod.

At the time this book was published, I had compiled 32.6 years of experience in four platforms: blogging, Twitter, Facebook, and Instagram.[18] I have worked online for nonprofit ministries for a decade, and never without an iPhone. And those labors have not insulated me from the pressing questions of the digital age—rather, they have amplified them. At the same time, my work has put me in contact with several of the most thoughtful Christian philosophers, theolo-gians, pastors, and artists who are thinking carefully about helping the church respond wisely to the digital age, and here I will share some of the best insights from my many conversations with them.

Simultaneously, I wrote this book in dialogue with a variety of Christians: students, singles, married couples, parents, homemak-ers, business professionals, and ministry leaders. Each of us faces

---

17. Tony Reinke, "Walk the Worldwide Garden: Protecting Your Home in the Digital Age," Desiring God, desiringGod.org (May 14, 2016).
18. I have been blogging for 565 weeks, posting on Twitter and Facebook each for 441 weeks, and using Instagram for 248 weeks.

similar questions about how to live healthy and balanced lives in the digital age.

BACKWARD DESIRES

Media ecologist Marshall McLuhan (1911–1980) reminded his generation that technology is always an extension of the self. A fork is simply an extension of my hand. My car is an extension of my arms and my feet, and no less so than Fred Flintstone's footmobile.

Likewise, my smartphone extends my cognitive functions.[19] The active neurons in my brain are a crackling tangle of skull lightning, and my thought life resembles a thunderstorm over Kansas.[20] This tiny electrical storm in the microscopic space of my nervous system quite naturally extends out to my thumbs to create tiny digital sparks of electricity inside my phone that beam out to the world by radio waves.

This all means that my phone marks a place in time and space— outside of me—where I can project my relationships, my longings, and the full scope of my conscious existence. In fact, hold up the word "desire" in a mirror and it will read "erised," the name of the magic mirror in the Harry Potter books.[21] In the ancient Mirror of Erised, you see the deepest longings of your heart revealed in vivid color. Our shiny smartphone screens do the same.

Too often what my phone exposes in me is not the holy desires of what *I know I should want*, not even what *I think I want*, and especially not what *I want you to think I want*. My phone screen divulges in razor-

---

19. "If the wheel is an extension of feet, and tools of hands, backs, and arms, then electromagnetism seems to be in its technological manifestations an extension of our nerves, and becomes mainly an information system." Marshall McLuhan, video interview, "The Future of Man in the Electric Age," marshallmcluhanspeaks.com (BBC, 1965). Throughout the book, I will distinguish between our lives as *embodied* and *disembodied*, not as precise terms but as useful terms of contrast. Of course, on our phones, we always use our bodies—our eyes, thumbs, ears, brains, and even our nerves to sense the phantom vibrations. The usefulness of the terms will become clear later in the book when we address the influence of our phones on our physical health, something we often ignore. They will also serve as a good contrast to the *embodied* life, a term I use in reference to scenarios in which all of our personhood—mind, body, soul, emotion—is displayed and used simultaneously (as in a face-to-face conversation).

20. A metaphor from N. D. Wilson's address, "Words Made Flesh: Stories Telling Stories and the Russian Dolls of Divine Creativity," Vimeo, vimeo.com (April 25, 2015).

21. J. K. Rowling, *Harry Potter and the Sorcerer's Stone* (New York: Scholastic, 1998), 207–8.

sharp pixels what my heart *really wants*.[22] The glowing screen on my phone projects into my eyes the desires and loves that live in the most abstract corners of my heart and soul, finding visible expression in pixels of images, video, and text for me to see and consume and type and share. This means that whatever happens on my smartphone, especially under the guise of anonymity, is the true exposé of my heart, reflected in full-color pixels back into my eyes.

Honestly, this may explain the passcodes. To get into a phone is to peek into the interior of another's soul, and we may be too ashamed for others to see what we clicked and opened and chased around online.

What could be more unsettling?

If we are honest enough to face our smartphone habits, and use the pages ahead as an invitation to commune with God, we can expect to find grace for our digital failures and for our digital futures. God loves us deeply, and he is eager to give us everything we need in the digital age. The spilled blood of his Son proves it.[23] We need his grace as we evaluate the place of smartphones—the pros and the cons—in the trajectory of our eternal lives. If we fluff it, not only will we suffer now, but generations after us will pay the price.

---

22. A haunting heart reality vividly described in James K. A. Smith, *You Are What You Love: The Spiritual Power of Habit* (Grand Rapids: Brazos, 2016), 27–38.

23. Rom. 8:32.

Introduction

A LITTLE THEOLOGY OF TECHNOLOGY

The moment when my first smartphone caught a wireless email outside that blustery rest stop in the Iowa cornfields is not where the story of this book begins. The launch of the iPhone at Macworld Expo 2007 is not far back enough either. Neither is the beginning of Apple or the birth of Steve Jobs. To see the timeline of the smartphone, we need a quick glance at the history of technology as it stretches back over the centuries. Our digital age is no cosmic accident.

THE STORY OF TECHNOLOGY

In the beginning, God created Adam out of mud and Eve out of a rib. Yahweh bent down and exhaled breath into their lungs, and they awoke into a strange world of oceans and sunshine and mountains and fruit and unnamed animals, untilled soil, and untapped materials, such as diamonds, gold, silver, and iron.[1] God first commanded his creatures to make babies, to collect food, and to govern the animals. But in those early commands, God already had drawn his endgame into his blueprints. The garden was only a beginning. The goal was a globe of technological advancement, leading to a creation so refined that the city streets will be paved thick with crystal gold, a creation

---

1. Gen. 2:10–14.

so radiant and luminescent that we can hardly imagine what it will look like in the end.[2] So when Adam and Eve awoke and walked into the garden, an unseen, much larger plan was also set in motion. The untilled garden would become a glorious city.

We find ourselves in the middle of this garden-to-city unfolding of history, and God is governing the entire process in several ways. Between the guardrails of natural law, as well as the guardrails of the abundance and scarcity of certain raw materials in the earth, and carried forward through his image bearers, each wired for innovation, the trajectory of technological progress—from the garden to the city—was set in motion. This process is entirely initiated, intended, and guided by God.[3]

But between the muddy rural beginning of the garden and the gleaming urban finale, we must fill in the story, because that's where we find ourselves: east of Eden, west of the Great City, journeying now in God's sovereignly guided history, holding smartphones. As the broader history of technology unfolds, the Bible teaches us nine key realities we must rehearse to ourselves in the digital age.

1. Technology modifies creation

God's commission to the first couple, to garden the globe and to raise animals, implied a series of technological advances that would make all of this work possible through stone tools, then copper tools, and then iron tools.

Unlike his other creatures, God's image bearers would grow food strategically. By design, agricultural advances began rather quickly— a trajectory of shovels, sickles, and horse-drawn plows, and then tractors, irrigation systems, and now GPS-guided (and GPS-driven!) equipment. Technology is used to subdue creation for human good, but also to increase efficiency. Today's agriculture is not perfect, and

---

2. Rev. 21:18–21.
3. This inevitability explains what historians call the phenomenon of "multiple discovery" or "simultaneous inventions." See Clive Thompson, *Smarter Than You Think: How Technology Is Changing Our Minds for the Better* (New York: Penguin, 2013), 58–66.

it raises moral questions, but the long train of technological advances here is especially illuminating and stunning.

Farming also is one example of technology built from the Creator's intelligence (given to mankind) and creation's abundance (supplied in the earth). Technology is the reordering of raw materials for human purposes. Adam and Eve reordered the raw materials of soil in order to make plants and flowers flourish. Today, chefs and cooks reorder the raw materials of foods into delicious meals. Framing carpenters reorder raw materials of lumber and nails to form homes. Pharmaceutical chemists reorder organic and synthetic elements into healing drugs. Musicians reorder notes and sounds into music. Novelists reorder the raw material of human experience into stories. As a writer of nonfiction, I reorder the raw materials of words and ideas for a publisher, which then reorders wood pulp, black ink, and binding glue into a book for you to hold and read. All of this is technology.

2. Technology pushes back the results of the fall

Not long into the story of the world, Adam and Eve made the tragic mistake—committing the inexplicable sin—of ignoring God's only prohibition. Satan tempted them, and Eve and Adam took a bite at becoming godlike. In that moment, God brought down his curse on creation, and the immediate result was a breakdown in man's relationships with everyone and everything.[4]

That breakdown still affects us today—weeds in the crops, pain in the delivery room, and embarrassment in nakedness. Farmers use weed-killing technology to minimize thorns and thistles on the farm. Women use pain-suppressing technology in childbirth. Fashion designers use fabric to cover our bodies. The sweep of technological advance is a gracious gift from God to help us live in a fallen creation. But all of this technology also reminds us of our fundamental problem—we are sinfully alienated from God.

---

4. Gen. 3:1–24.

3. Technology establishes human power

Unhitched from fear and obedience to God, technology quickly becomes a pawn in human power plays. The discovery of copper and the invention of stronger and harder carburized iron brought easier farming, but it also brought new equipment for warfare.[5] To own iron mines and employ blacksmiths was to control an endless supply of new weaponry, and to control an endless supply of new weaponry was to flex military superiority, and to flex military superiority was to wield power over rival nations. Bows, arrows, iron, and gunpowder all give power to defend and conquer. The same holds true today. Power and superiority rest on technology: atomic weapons, warships, drones, fighter jets, and missiles. The larger a nation's military, the more power it can wield in the world. Such a quantifiable and scalable power is possible only through technological innovation.

4. Technology helps to edify souls

In the biblical storyline, innovations also serve worshipers.

Musical instruments were invented in order for God's people to express their joy in beautiful songs.[6] Later, the temple of Israel exhibited years of advances in building technology, metallurgy, and artistic craftsmanship. The greatness and the majestic scale of the temple proclaimed to the nations the glory, greatness, and splendor of Israel's God.

As God's plan moved from a come-and-see religion (Old Testament) to a go-and-tell focus (New Testament), chisel and stone gave way to primitive advances in paper and ink, making it possible for written communications technology to advance. God's words, first scratched in stone, then on processed animal skins, and then on products of trees, would become the Creator's centerpiece for drawing together his people separated by continents, languages, and millennia. Over time, the many scrolls of the Old Testament and the many books and letters of the New Testament were gathered into a codex,

---

5. Gen. 49:5; Judg. 1:19; 4:3.
6. 1 Chron. 15:16; 23:5.

translated, and mass-published as a single book of unified authority that we now conveniently carry in one hand. Every time we open our Bibles, our souls are being fed through centuries of technological advancement.

From trumpets and temples to gold-edged Bibles, God intended technology to play an essential role for us to know and worship him.

5. Technology upholds and empowers our bodies

Technological advances change and refine our bodies in very dramatic ways, too. Eyeglasses and hearing aids boost our senses of seeing and hearing. Musical technology, such as the violin, fine-tunes human motor skills and gives us new purposes for the microrefined movements of our bodies. Industrial technology connects our hands to the hydraulic arms of digging machines. Medical technology starts stopped hearts and sustains dying bodies. Advances in medicine cure diseases and slow terminal illnesses. And advances in clothing make it possible for us to adorn our bodies in ways that define and shape the identities we project to one another.[7]

Technology enhances our bodies, refines our movements, amplifies our actions, and shapes how we present ourselves to the world.

6. Technology gives voice to human autonomy

The good-bad-ugly mix of technology came to a particularly obnoxious expression at the Tower of Babel, an attempt to consolidate all known building innovation to build a rebel city.[8] More than a simple skyscraper, Babel was a new empire with a central city unified around a temple (the tower), all dedicated to the worship of human progress. Suppressing God's ingenuity in all human advances, Babel was man's attempt to hijack technology and to fabricate an entire society and religious life in rebellion to the Creator.

As such, Babel marked man's collective rejection of the idea that technology is a gift from God. Before they built a tower into the sky,

---

7. 1 Pet. 3:3–4; 1 Tim. 2:9; Rev. 17:4–5.
8. Gen. 11:1–9.

the people of Babel drew a line in the sand that said to the Creator, "Human autonomy will take credit for technological innovation from here on, *thankyouverymuch.*" The mockery of this treasonous act is also partially comic—man builds his temple *up* as high as possible, and then the living God of the universe stoops *down* to his knees and puts his cheek on the ground in order to evaluate the progress.[9] This is always what happens when technology is misused in unbelief. God is the genesis of all knowledge and technological advance, and he is the author and finisher of a glorified city to come. Why would a mud skyscraper impress him?

Technology is not inherently evil, but it tends to become the platform of choice to express the fantasy of human autonomy.

7. God governs every human technology

The Tower of Babel was really the Tower of Ignorance. This skyscraper of pride was assembled with earth's raw materials and shaped by human ingenuity—and all of these gifts came from God. To build a godless skyscraper, using God's resources put in the ground and God's inventiveness put in his image bearers, was the height of human arrogance and (as we will see later) the total distortion of human purpose.

So God scattered the builders across the globe by a variety of languages (and drew all those languages back together at Pentecost when the gospel was ready for worldwide distribution[10]). God was not absent at Babel. He was the cosmic foreman on site, overruling human technology to serve his ultimate gospel purpose.

But God's sovereign reign over the most horrific evils of technology is nowhere clearer than in the Roman cross. An upright wooden post with a transverse beam, the cross was a showcase for a criminal: nailed down by three iron spikes, he was then lifted up for all to see as the cross was planted in the ground. The cross was designed to kill criminals, insurrectionists, and disobedient slaves, and to do so

---

9. Gen. 11:5.
10. Acts 2:1–13.

slowly by exhaustion and asphyxiation. The slow death was public torture, a billboard of intimidation: Behold the fate of any fool who defies Roman rule and threatens social stability.[11]

But this awful tool of torture doubled as the hinge on which all of God's redemptive plan turned. God created trees to serve man, but man invented crosses to destroy man. In the darkness of this most evil moment, God's entire plan for the glorious new city took a decisive step forward. Through an evil misuse of technology, man killed the Author of life, yet God was sovereign over the entire process.[12] By a cosmic paradox that will never be eclipsed, in the naked torture of shame before the eyes of man, Christ exposed all the forces of evil to the shame of stripped-naked defeat.[13]

Evil was defeated by technology, all by God's sovereign design. Technology, even in the hands of the most evil intention of man, is never outside the overruling plan of God. In this case, Calvary was *hacked*. God broke into the technology of the cross "and with a little twist reversed its function."[14] God does this: he makes a mockery of our evil technologies through his sovereign hackery.

8. Technology shapes every relationship

The lineage of technological advance is long—bows and arrows, wheels and axles, iron tools and weapons, movable type and printing presses, clocks and watches, steam engines and railroads, cars and jets, computers and smartphones. Every new technology opens humanity to new hopes, dreams, and aspirations. Every technology changes the fundamental social dynamics of how we relate to the world, to one another, and to God.

First, technology changes how we relate to the earth. With a GPS app, I can see my exact place on the earth in a way that was almost impossible twenty years ago and unfathomable to my ancestors.

---

11. Martin Hengle, *Crucifixion* (Minneapolis: Fortress Press, 1977).
12. Acts 3:15; 2:23.
13. Col. 2:15.
14. Martin M. Olmos, "God, the Hacker: Technology, Mockery, and the Cross," *Second Nature*, secondnaturejournal.com (July 29, 2013).

Second, technology changes the way we relate to one another. If I approach you on the street and begin chatting, our relationship is fundamentally open. But if I approach you for a chat and my video recorder app is open and I am holding my phone out in front of me, our interaction is fundamentally changed as you try to decide if you will make eye contact with me or with the invisible audience watching on the other side of my mini camera lens.

Third, technology can become a metaphor that God uses to reveal his work in the world. Once we had made primitive advancements in metallurgy, for example, God could reveal his work in humanity as a consuming fire who smelts mankind—to judge the dross of rebellion and to purify his handiwork, his nation, of false alloys. The unveiling of new technology creates new metaphors for God to reveal how he engages with us mortals.[15]

9. Technology shapes our theology

Finally, we use technology to manifest metaphors of God (for good or ill). Take the more recent technology of the pocket watch—miniature hairsprings, winding wheels, and precise gears, all wound up into rhythmic clicking. With the invention of the watch, we could keep time with accuracy and choreograph our schedules. The technological advance in timepieces also birthed two new metaphors to explain God's relationship to us—one perceptive, the other deceptive.

First, the watch provided a helpful metaphor for God. Since the watch's various pieces all come together to serve one function in the end, it bears all the marks of "intelligent design," the handiwork of one designer. Such is also true of our bodies. Together, the various parts and pieces and chemicals of our existence join in harmony to sustain our cohesive existence. This is "the watchmaker analogy." God is not only close; his fingerprints are on us.

---

15. Isa. 1:22–25; Jer. 6:27–30; Ps. 119:119. See also Paula McNutt, *The Forging of Israel: Iron Technology, Symbolism and Tradition in Ancient Society* (Sheffield, England: Bloomsbury T&T Clark: 2009). It should be said that God coined new metaphors of technology for himself until the closing of the canon.

But the watch also provided a faulty metaphor for God. Some began to imagine a God who assembled the universe, wound it up, set it in motion, and walked away. This is a form of deism, the idea that God is generally withdrawn and remote from the world apart from preserving natural laws.

For better or worse, technology fundamentally changes how we talk about God. And technology shapes the way God communicates himself to us. God makes himself clear to us through metaphors of technology, and we find it possible to define him, and also to distort him, by projecting metaphors of technology onto him.

TECHNOLOGY THEOLOGY

I've only skimmed the depths here. My point is that every technological innovation is a new theological invitation for renewed biblical contemplation by God's people. That means several things.

First, life in the digital age is an open invitation for clear, biblical thinking about the impact of our phones on ourselves, on our creation, on our neighbors, and on our relationships to God. Thoughtlessly adopting new technology is worldliness.

Second, technology is technology, whether tethered to an outlet or to a horse. For this project, I will not make a hard-and-fast distinction between *tools* and *technology*, disconnecting primitive tools off the electrical grid from newer technologies we plug in. Partly this is because household gods of carved stone or wood and handheld idols of silver and gold, common in the ancient world, were not tools. These idols were more like our technologies, divine oracles of knowledge and prosperity, used by worshipers in an attempt to control and manipulate the events of life for personal benefit. The figurine and the iPhone appeal to the same fetish.

Third, whatever my smartphone is doing to me, it is also pointing me toward a glorious city to come. We do not trust in handheld things. We do not trust in handmade things. Instead, we long to be in the presence of our triune God in a new creation, built not by human ingenuity and sinful hands, but by the very design and innovation of God—the sinless and deathless and tearless creation God has always intended.[16]

OUR PLACE IN HISTORY

So here we are, in "the digital age," an age so thick with innovation that we grow blind to it. And we are adopting and adapting to new technologies faster than any generation in world history. As of 2015, among American adults eighteen to twenty-nine years old, 86 percent own a smartphone, up from 52 percent four years earlier. In the same demographic, 50 percent own a tablet, up from just 13 percent four years earlier. Concurrently, among the same demographic, ownership of computers, MP3 players, game consoles, and ebook readers declined.[17] Our phones are gobbling up these functions.

Perhaps we adapt so readily because we are a gifted generation, easily trainable and moldable. Or perhaps we adapt so readily because, as Jacques Ellul suggested, our technology exerts a sort of terrorism over us.[18] We live under the threat that if we fail to embrace new technologies, we will be pushed aside into cultural obsolescence, left without key skills we need to get a job, disconnected from cultural conversations, and separated from our friends.

Whatever our motives, the fact remains—we are adopting, we are going online, and we are going mobile. Smartphone cases double as wallets because we wouldn't dare leave the house without them. In fact, 36 percent of eighteen- to twenty-nine-year-olds in America admit they are online "almost constantly"—a phenomenon made possible by the smartphone. The most likely adult to live online makes more than $75,000 per year, is a college graduate, lives in a nonrural setting, and is in the eighteen-to-twenty-nine age range.[19] Our mobile web addiction may be new, but it's here to stay. We are never offline.

---

16. John 14:1–7; Acts 7:49–50; Heb. 9:11–28.
17. Monica Anderson, "Technology Device Ownership: 2015," Pew Research Center, pew internet.org (Oct. 29, 2015).
18. Jacques Ellul, *The Technological Bluff* (Grand Rapids, MI: Eerdmans, 1990), 384–400.
19. Andrew Perrin, "One-Fifth of Americans Report Going Online 'Almost Constantly,'" Pew Research Center, pewinternet.org (Dec. 8, 2015).

So is my smartphone a hostile enemy? Is it a cultural trinket? Is it a legitimate tool? Those are a few of the questions we will examine in the pages ahead. Our phones have concentrated powerful technology into a little device we control with our thumbs. We have full access to this technology, and by some kind of digital and electrical magic, we are potentially connected at all times with every other phone on the planet.

All of these realities *are* changing us; there's no debate on that. The bigger questions remain: *How* are our smartphones changing us? And should we be concerned?

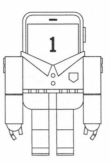

WE ARE ADDICTED TO DISTRACTION

We check our smartphones about 81,500 times each year, or once every 4.3 minutes of our waking lives, which means you will be tempted to check your phone three times before you finish this chapter.[1]

The impulse is not hard to understand. Our lives are consolidated on our phones: our calendars, our cameras, our pictures, our work, our workouts, our reading, our writing, our credit cards, our maps, our news, our weather, our email, our shopping—all of it can be managed with state-of-the-art apps in powerful little devices we carry everywhere. Even the GPS app on my phone, which guided me to a new coffee shop today, possesses thirty thousand times the processing speed of the seventy-pound onboard navigational computer that guided Apollo 11 to the surface of the moon.

It's no wonder we habitually grab our phones first thing in the morning, not only to turn off our alarms, but also to check email and social media in a half-conscious state of sleep inertia before our groggy eyes can fully open. If the ever-expanding universe is humankind's final horizon outward, our phones take us on a limitless voyage inward, and we restart the journey early every morning.

---

1. Jacob Weisberg, "We Are Hopelessly Hooked," *The New York Review of Books* (Feb. 25, 2016).

I am no stranger to this instinctive phone grab, but I wanted to see if others shared this pattern, so I surveyed eight thousand Christians about social-media routines.[2] More than half of the respondents (54 percent) admitted to checking a smartphone within minutes of waking. When asked whether they were more likely to check email and social media *before* or *after* spiritual disciplines on a typical morning, 73 percent said *before*. This reality is especially concerning if the morning is when we prepare our hearts spiritually for the day. (We will look more closely at this habit, and my other findings, in the chapters ahead.)

Our phones are addictive, and, like addicts, we seek hits immediately in the morning. And, yes, there's an app for that.

FACEBOOK

The app we most often turn to for our hits is Facebook. In 2013, 63 percent of Facebook users checked in daily. Just one year later, that number had shot up to 70 percent. If you check Facebook every day, you join more than one billion others with the same compulsive routine. And the average user now spends fifty minutes—every day—in the Facebook product line (Facebook, Messenger, Instagram), a number that continues to surge by strategic design.[3]

The Facebook uptick coincides with a spike in mobile technology and a spike in users who are adopting smartphones into every open moment of their lives. Facebook now travels with us, and this mobility is quickly making Facebook addicts of us all. Few of us can stop ourselves. Ofir Turel, a psychologist at California State University-Fullerton, warns that Facebook addicts, unlike compulsive drug abusers, "have the ability to control their behavior, but they don't have the motivation to control this behavior because they don't see the consequences to be that severe."[4]

---

2. This was a nonscientific survey of desiringGod.org readers conducted online via social-media channels (April 2015). I will return to our findings later in the book.

3. James Stewart, "Facebook Has 50 Minutes of Your Time Each Day. It Wants More," *The New York Times* (May 5, 2016).

4. Rebecca Strong, "Brain Scans Show How Facebook and Cocaine Addictions Are the Same," BostInno, bostinno.streetwise.co (Feb. 3, 2015).

But the consequences are real. As digital distractions intrude into our lives at an unprecedented rate, behavioral scientists and psychologists offer statistical proof in study after study: the more addicted you become to your phone, the more prone you are to depression and anxiety, and the less able you are to concentrate at work and sleep at night. Digital distractions are no game. Because we are all so interconnected, hundreds of people (friends, family members, and strangers) can interrupt us at any moment. And when we are bored, with the flick of a thumb we can skim an endless list of amusements and oddities online.

The psychological and physical consequences of our digital distractions are interesting, but this book will instead focus on the spiritual dimensions of our smartphone addictions—consequences almost entirely ignored in many Christian articles and books. As we progress, I will point out some scientific findings, but only as a turnstile for us to move the discussion from the biological effects of our screen habits into the more important discussion of the spiritual push and pull between our online actions and the infinite consequences of our device behaviors. Eternity, not psychology, is my deepest concern.

So if the study of online trends shows a tsunami of digital distractions crashing into our lives, we need situational wisdom to answer three spiritual questions: Why are we lured to distractions? What is a distraction? And, most foundational of all, what is the undistracted life?

WHY DISTRACTIONS LURE US

Unhealthy digital addictions flourish because we fail to see the consequences, so let's begin our study by uncovering three reasons why we succumb to distractions so easily.

First, we use digital distractions to keep work away. Facebook is a way of escape from our vocational pressures. We procrastinate around hard things: work deadlines, tough conversations, laundry piles, and school projects and papers. The average American college student wastes 20 percent of class time tinkering on a digital device, doing things unrelated to class (a statistic that seems low to me!).[5] When life becomes most demanding, we crave something else—anything else.

Second, we use digital distractions to keep people away. God has called us to love our neighbors, yet we turn to our phones to withdraw from our neighbors and to let everyone know we'd rather be somewhere else. In a meeting or a classroom, if my phone is put away, I am more likely to be perceived as engaged. If my phone is not in use, but is faceup on the table, I present myself as engaged for the moment, but possibly disengaged if someone more important outside the room needs me. And if my phone is in my hand, and I am responding to texts and scrolling social media, I project open dismissiveness, because "dividing attention is a typical expression of disdain."[6]

In the digital age, we are especially slow to "associate with the lowly" around us.[7] Instead, we retreat into our phones—projecting our scorn for complex situations or for boring people. In both cases, when we grab our phones, we air our sense of superiority to others—often without knowing it.

Third, we use digital distractions to keep thoughts of eternity away. Perhaps most subtly, we find it easy to fall into the trap of digital distractions because, in the most alluring new apps, we find a welcome escape from our truest, rawest, and most honest self-perceptions. This was the insight of seventeenth-century Christian, mathematician, and proverb-making sage Blaise Pascal. When observing distracted souls of his own day (not unlike those of our time), he noticed that if you "take away their diversion, you will see them dried up with weariness," because it is to be ushered into unhappiness "as soon as we are reduced to thinking of self, and have no diversion."[8] Pascal's point is a perennial fact: the human appetite for distraction is high in

---

5. Leslie Reed, "Digital Distraction in Class Is on the Rise," Nebraska Today, news.unl.edu (Jan. 15, 2016).

6. Oliver O'Donovan, *Ethics as Theology*, vol. 2, *Finding and Seeking* (Grand Rapids, MI: Eerdmans, 2014), 45.

7. Rom. 12:16.

8. Blaise Pascal, *Thoughts, Letters, and Minor Works*, ed. Charles W. Eliot, trans. W. F. Trotter, M. L. Booth, and O. W. Wight (New York: P. F. Collier & Son, 1910), 63.

every age, because distractions give us easy escape from the silence and solitude whereby we become acquainted with our finitude, our inescapable mortality, and the distance of God from all our desires, hopes, and pleasures.

Driving every diversion, from international warfare to international tourism, is the promise of escaping boredom at home, said Pascal in his day: "I have discovered that all the unhappiness of men arises from one single fact, that they cannot stay quietly in their own chamber."[9] Staring at the ceilings of our quiet bedrooms, with only our thoughts about ourselves, reality, and God, is unbearable. "Hence it comes that men so much love noise and stir; hence it comes that the prison is so horrible a punishment; hence it comes that the pleasure of solitude is a thing incomprehensible."[10] To be without the constant availability of distraction is solitary confinement, a punishment to be most dreaded. That is why in those moments when we realize we have forgotten our phone, lost it, or let the battery run out, we taste the captivity of a prison cell, and it can be frightening.

Although we have a thousand reasons to be sobered by our self-reflection, we seek amusements, like "playing billiards or hitting a ball,"[11] or, for us, downloading a new ninety-nine-cent game. Our ever-present phones offer endless diversions, from ten-second downloads to one-touch purchases. Our pings, alerts, and push notifications all redirect us from our greatest needs and realities.

The Pascal of our generation puts it this way: "We run away like conscientious little bugs, scared rabbits, dancing attendance on our machines, our slaves, our masters"—clicking, scrolling, tapping, liking, sharing . . . anything. "We think we want peace and silence and freedom and leisure, but deep down we know that this would be unendurable to us." In fact, "we *want* to complexify our lives. We don't *have* to, we *want* to. We want to be harried and hassled and busy. Unconsciously, we want the very thing we complain about. For if we

---

9. Ibid., 52.
10. Ibid., 53.
11. Ibid., 55.

had leisure, we would look at ourselves and listen to our hearts and see the great gaping hole in our hearts and be terrified, because that hole is so big that nothing but God can fill it."[12]

To numb the sting of this emptiness, we turn to the "new and powerful antidepressants of a non-pharmaceutical variety"—our smartphones.[13] But even as we seek escape in social media, death follows us and haunts those digital diversions in new ways. "I love the fun and frivolity of much of Twitter. The GIFs. The jokes. The nested conversations," admits one honest writer. "The reality is, though, deep down there's part of me that's scared that if I'm out of sight, I'll be out of mind, and I won't matter anymore. In a sense, this is one dimension of the looming fear of death that most of us in contemporary American society never want to wrestle with or name anymore."[14] No, we don't. All of us find ourselves uncomfortably close to passing into the mystery of eternity, leaving this place, and being forgotten in the only home we've ever known. So every day we jump back into the hamster wheel of our digital conversations and muffle the reality.

The philosophical maxim, "*I think*, therefore I am,"[15] has been replaced with a digital motto, "*I connect*, therefore I am,"[16] leading to a status desire: "*I am 'liked,'* therefore I am."[17] But our digital connections and ticks of approval are flickering pixels that cannot ground the meaning of our lives. And yet, I seek to satisfy this desire every time I cozy up to the Facebook barstool, to be where every friend knows my name, where my presence can be affirmed and reaffirmed at virtual points throughout the day. I want anything to break the silence that makes me feel the weight of my mortality.

So here's an exercise to help ground our self-perception. Once a

12. Peter Kreeft, *Christianity for Modern Pagans: Pascal's Pensées Edited, Outlined, and Explained* (San Francisco: Ignatius, 1993), 168–69.

13. Andrew Sullivan, "I Used to Be a Human Being," *New York* magazine (Sept. 18, 2016).

14. Derek Rishmawy, "Forget Me Not (Twitter and the Fear of Death)," *Reformedish*, derek zrishmawy.com (April 6, 2016).

15. René Descartes, *The Philosophical Works of Descarte*, trans. E. S. Haldane and G. R. T. Ross (New York: Cambridge University Press, 1970), 101.

16. Kevin Vanhoozer, interview with the author via email (Feb. 26, 2016).

17. Donna Freitas, *The Happiness Effect: How Social Media Is Driving a Generation to Appear Perfect at Any Cost* (New York: Oxford University Press, 2017), 33.

day, set your phone down for a moment, hold out your right hand, palm out and fingers to the sky, and imagine the timeline of history reaching a mile to your left and an eternity to your right. Your time on earth intersects roughly the width of your hand (give or take).[18] Nothing puts social media and smartphone habits into context like the blunt reality of our mortality. Let it sink in a bit. Feel the brevity of life, and it will make you fully alive.[19]

DEFINING DISTRACTIONS

This is all pretty heavy, I know, but if we are honest, we need a dose of Pascal's prophetic warnings today. "We live in a very loquacious, noisy, distracted culture," says philosopher Douglas Groothuis, who has been tracking the digital world's influence on Christians for more than twenty years since writing his 1997 book, *The Soul in Cyberspace.* "It is difficult to serve God with our heart, soul, strength and mind when we are diverted and distracted and multi-tasking everything."[20] Historian Bruce Hindmarsh adds, "Our spiritual condition today is one of spiritual ADD."[21]

If Pascal sounds like he has taken the discussion too far, in reality he hasn't taken it far enough. His warnings about the distractions of untimely amusements only mimic the urgency of the biblical warnings on distractions, which further broaden the categories until "distraction" covers all of the immediately pressing details of our daily lives, relationships, and apparent duties, and even our pursuits of money and possessions—anything that preoccupies our attention on this world and life. A distraction can come in many forms: a new amusement, a persistent worry, or a vain aspiration. It is something that diverts our minds and hearts from what is most significant; anything "which monopolizes the heart's concerns."[22] The heart works best when it is not dominated by cares and demands.

---

18. Ps. 39:4–5.
19. Ps. 90:12.
20. Douglas Groothuis, interview with the author via phone (July 3, 2014).
21. Bruce Hindmarsh, interview with the author via phone (March 12, 2015).
22. Horst Robert Balz and Gerhard Schneider, *Exegetical Dictionary of the New Testament* (Grand Rapids, MI: Eerdmans, 1990), 2:409.

In six places, the New Testament warns us about the effects of unchecked distractions on the soul, and we can boil those distractions down into three potent categories:

1. *Unchecked distractions that blind souls from God.* These are the most dangerous distractions: worldly worries, anxieties, and pursuits of wealth, self-centered concerns with personal security that suffocate the soul by snatching away seeds of truth, choking off the fruit of the gospel, and rendering its hope irrelevant. The vanity of the ephemeral robs our lives of what has infinite value.[23]

2. *Unchecked distractions that close off communion with God.* These distractions are exemplified in Martha, who was so distracted by her table service for others that she missed the importance of Christ's words for her own life.[24] We can become so unfocused in life that we get lost in the unforgiving wheel of daily tasks and fail to listen to the voice of Christ. We fail to pray and fail to see him as intently listening and drawing near to us. God feels distant because we are distracted. Yet he seeks us; he seeks our undivided attention.[25]

3. *Unchecked distractions that mute the urgency of God.* Marriage is a beautiful gift, but it also comes packaged with routines and obligations—certain domestic *distractions*—demanding much attention. In embracing the blessings of marriage, spouses also willingly accept the distractions of the married life and relinquish what Paul sees as the "undistracted" life—the gift of singleness.[26]

Marriage is not the ultimate priority of life; neither is romantic love or sex. Marriage is a precious gift, and intimacy in marriage is a beautiful expression of God's design—but Scripture calls for seasons when even sex should cease in order for spouses to recalibrate their prayer lives and to reset their greater priority of communion with God.[27]

Marriage and singleness are both profound gifts. Marriage affirms the goodness of creation,[28] projects a beautiful metaphor of

---

23. Matt. 13:22; Mark 4:19; Luke 8:14.
24. Luke 10:38–42.
25. Luke 21:34–36.
26. 1 Cor. 7:32–35.
27. 1 Cor. 7:1–5.
28. Matt. 19:4–6; 1 Tim. 4:1–5.

Christ's love for his church,[29] and anticipates a cosmic marriage to come.[30] Singleness, on the other hand, points our attention back to the beautiful life of Christ on earth and forward to the majesty of our soon-coming moment of personal glorification.[31] Foreshadowing that moment of metamorphosis, Christ pictures a singleness so profound and regal that all earthly singleness finds transcendent urgency and unquestionable dignity. In each case, marriage and singleness are divine gifts, validated by Christ, celebrated by Paul.

First Corinthians 7 is the most detailed biblical theology of *distraction* and the pursuit of *undistraction*. Once we wrestle through what it means for marriage, we are positioned to apply those same categories to our digital lives. True distractions include anything (even a good thing) that veils our spiritual eyes from the shortness of time and from the urgency of the season of heightened expectation as we await the summing up of all history.

The date of Christ's return is a secret, but it approaches so rapidly that it calls for every Christian to remain on his or her toes in anticipation.[32] The death and resurrection of Christ has marked the beginning of the end, the runoff, the moment when a soccer match clock exceeds ninety minutes and keeps ticking for some amount of unknown stoppage time, soon to finally expire. The clock on God's redemptive timeline is past ninety minutes, and ticking. From now on, whenever we attempt to define *distractions*, especially in the most complex areas of life—such as dating, sex, and marriage—we must seek to do so by seeing ourselves inside of God's urgent and soon-to-end timeline for this creation.

All distractions are measured by the reality that "the appointed time has grown very short."[33] We are called to *watchfulness*[34] because

---

29. Eph. 5:22–33.
30. Rev. 19:6–10.
31. Mark 12:25; 1 Cor. 7:29. In the complex questions about marriage, divorce, and singleness in 1 Corinthians 7, answers must be "worked out in the context of the priorities of the gospel and the transformed vision brought about by the dawning of the eschatological age and the anticipation of the end." D. A. Carson, sermon, "The Gospel of Jesus Christ; 1 Cor. 15:1–19," The Gospel Coalition, thegospelcoalition.org (May 23, 2007).
32. Matt. 24:36–25:13; 1 Thess. 5:1–11.
33. 1 Cor. 7:29.
34. Matt. 24:42; 1 Cor. 16:13; Col. 4:2.

everything in the Christian life is conditioned by this sense of the eschatological urgency of Christ's return.[35] For those with eyes to see, Christ's return is so imminent, it potently declutters our lives of everything that is superficial and renders all of our vain distractions irrelevant. To put it another way, our battle against the encumbering distractions of this world—especially the unnecessary distractions of our phones—is a heart war we can wage only if our affections are locked firmly on the glory of Christ. The answer to our hyperkinetic digital world of diversions is the soul-calming sedative of Christ's splendor, beheld with the mind and enjoyed by the soul. The beauty of Christ calms us and roots our deepest longings in eternal hopes that are far beyond what our smartphones can ever hope to deliver.[36]

THE UNDISTRACTED LIFE?

So should we turn back the clock and return to the simplicity of the "distraction-free" predigital age? No—there may have been a pre-digital age, but there has never existed a life without distractions. Whether you have a smartphone, a dumb phone, or no phone, you cannot escape a life that divides your attention. However, the Bible makes clear that those distractions fall on a spectrum. We face sancti-fied distractions and unsanctified distractions. We face soul-filling distractions and soul-deadening distractions. We face necessary interruptions and worldly interruptions. We face unavoidable dis-tractions of godly marriage and avoidable distractions of consumer culture. From the outset of this study, we must die to the idea that a distraction-free life is possible—it is not, and it never has been. The holy life is piously complex, meaning we must learn how to apply distraction management in every situation.

Here's the warning: as Christians, if we fail to manage life's dis-

---

35. Rom. 13:11–14.

36. See John Owen, *Meditations and Discourses on the Glory of Christ*, in The Works of John Owen, ed. William H. Goold (Edinburgh: Banner of Truth Trust, 1965), 1:277–79, 402–3. In this life, where we so often struggle with self-love, worldliness, endless cares and fears, and with "an excessive valuation of relations"—think: social media—in contrast, our souls must be fed "sedate meditations on Christ and his glory" (1:403).

tractions wisely, we will lose our urgency and—in the sobering words of one smartphone-addicted mom of young children—we may "forget how to walk with the Lord."[37] Distraction management is a critical skill for spiritual health, and no less in the digital age. But if we merely exorcise one digital distraction from our lives without replacing it with a newer and healthier habit, seven more digital distractions will take its place.[38] Over time, we may lose our hearts by the erosive power of unchecked amusements. Eventually we ignore Paul as we lose a sense of our place in God's timeline.

UNDISTRACTED ON PURPOSE

While our relationships with our phones may not be lifelong covenant relationships (though carrier contracts can feel like it), I would not be the first to suggest that owning a smartphone is similar to dating a high-maintenance, attention-starved partner.[39] The smartphone is loaded with prompts, beeps, and allurements. Many of these stimuli (perhaps most of them) are not sinful, but they are pervasive.

The more distracted we are digitally, the more displaced we become spiritually. Following Paul's words to married couples, we must make it our aim to purge our lives of all unnecessary and unhelpful distractions. Pastor Tim Keller was once asked online: Why do you think young Christian adults struggle most deeply with God as a personal reality in their lives? He replied: "Noise and distraction. It is easier to tweet than pray!"[40] (Said on Twitter, no less!) The ease and immediacy of Twitter is no match for the patient labor of prayer, and the neglect of prayer makes God feel distant in our lives.

As in every age, God calls his children to stop, study what captures their attention in this world, weigh the consequences, and fight for

37. Tracy Fruehauf, "Airing My Dirty Laundry," *One Frue Over the Cuckoo's Nest*, onefrueover thecuckoosnest.com (Aug. 18, 2015).

38. Matt. 12:43–45; Luke 11:24–26.

39. Trip Lee, interview with the author via Skype, explaining his track "iLove" (March 25, 2015). The same metaphor appears in Freitas, *The Happiness Effect*, 224.

40. Tim Keller (@timkellernyc), Twitter, twitter.com (Dec. 31, 2013).

undistracted hearts before him. To that end, here are ten diagnostic questions we can ask ourselves in the digital age:

1. Do my smartphone habits expose an underlying addiction to untimely amusements?
2. Do my smartphone habits reveal a compulsive desire to be seen and affirmed?
3. Do my smartphone habits distract me from genuine communion with God?
4. Do my smartphone habits provide an easy escape from sobered thinking about my death, the return of Christ, and eternal realities?
5. Do my smartphone habits preoccupy me with the pursuit of worldly success?
6. Do my smartphone habits mute the sporadic leading of God's Spirit in my life?
7. Do my smartphone habits preoccupy me with dating and romance?
8. Do my smartphone habits build up Christians and my local church?
9. Do my smartphone habits center on what is necessary to me and beneficial to others?
10. Do my smartphone habits disengage me from the needs of the neighbors God has placed right in front of me?

Let's be honest: our digital addictions (if we can call them that) are welcomed addictions. The key is to move from being distracted on purpose to being less and less distracted with an eternal purpose. The questions sting, and they touch every area of life—God, spouse, family, friends, work, leisure, and self-projection. But this sting can lead us to make healthy changes.

Our smartphones amplify the most unnecessary distractions as they deaden us to the most significant and important "distractions," the true needs of our families and neighbors. My phone conditions me to be a passive observer. My phone can connect me to many friends, but it can also decouple me from an expectation for real-life engagement. When I go into my social media streams, too often I use Facebook to insulate me from the real needs of my friends. Facebook becomes a safe and sanitized room where I can watch the ups and downs of others as an anonymous spectator, with no compulsive impulse to respond and care in any meaningful way. And as I do, I become more and more blind to the flesh and blood around me. That change is next on the list.

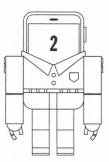

WE IGNORE OUR FLESH AND BLOOD

We know we should not neglect others, but we ignore our consciences and do it anyway. This neglect takes on a most dangerous form in the phenomenon of distracted driving.

Texting and driving is such a commonplace habit, the stats are now canonical. *Talking* on the phone while driving a vehicle makes you four times more likely to get into an accident, but *texting* while driving makes your chance of a crash *twenty-three times more likely*. Assuming a driver never looks up in the average time it takes to send a text (4.6 seconds), at fifty-five miles per hour, he drives blindly the length of a football field. Texting and driving is so idiotic, forty-six of fifty states have banned it.

But even these frequently cited facts haven't brought a stop to this drastically reckless distraction. They've hardly made a dent. Likewise, the laws against texting and driving have had little impact. One study by the University of Michigan concluded that anti-texting-and-driving laws might actually be causing *a rise* in the most serious texting-and-driving accidents.[1]

---

1. Johnathon P. Ehsani, C. Raymond Bingham, Edward Ionides, and David Childers, "The Impact of Michigan's Text Messaging Restriction on Motor Vehicle Crashes," *Journal of Adolescent Health* (Jan. 3, 2014).

WHY THE LAWS DON'T WORK

Why don't the laws work? And why are the deadliest texting-and-driving accidents on the rise?

Journalist Matt Richtel wrote *A Deadly Wandering* to answer these questions after investigating a 2006 crash caused by a college student who was texting and driving when his car swerved and collided with an oncoming vehicle, killing two people.[2] He retells the tragic accident, follows the consequent trial, and asks relevant questions about our legal obligations for maintaining undivided concentration in a digital world.

In the end, Richtel points one finger of blame for distracted driving in the direction of telecommunications marketers. We are fed mixed messages, he says. For example, in 2013, telecommunications giant AT&T released the commercial "Dizzy," a thirty-second spot featuring four young kids at a table answering a lone question from the moderator. "What's better?" he asks, "Doing two things at once or just one thing at once?" Of course, the children yell out the obvious answer: "Two." It's not complicated, we are told. Even little kids know it's better to do two things at once.

At the same time, AT&T was also funding famed documentarian Werner Herzog's anti-texting-and-driving film, *From One Second to the Next*, as part of AT&T's impressive "It Can Wait" campaign and website. Nearly eight million drivers have taken the online pledge "to keep your eyes on the road, not on your phone."[3]

So we must ask: Is accomplishing two things at once really a no-brain default answer that any child can come up with? No, it's not that simple.

But I think there's an even simpler explanation for why the laws don't work. As any high school teacher can tell you, we are inventive creatures when it comes to covert use of our phones. Laws banning texting are nearly unenforceable, but the states that crack down the

---

2. Matt Richtel, *A Deadly Wandering: A Mystery, a Landmark Investigation, and the Astonishing Science of Attention in the Digital Age* (New York: William Morrow, 2015).

3. See itcanwait.com.

hardest only make the practice more clandestine. In a car, you can send texts with one thumb under the window-level view of onlookers. The harder police clamp down on texting, the *lower* the phones go, and the lower the phones, the further drivers' attention is drawn off the road, requiring slightly more time for them to read and send texts, and more time to reorient their attention to their driving. Thus, the harder the attempt to stop texting and driving, the more concealed (and dangerous) texting becomes, and the more serious the accidents that result.

If laws, police enforcement, and fines cannot stop texting and driving, the solution must be bloody—and it is. Graphic ad campaigns show just how fast a careless driver can text and drive unspeakable destruction into the lives of others in oncoming traffic. Public service announcements reenact collisions in slow motion, with the shattering of glass, crumpling of metal, and tossing of human bodies. Those ads tap the real cause of texting and driving—a lack of awareness of the flesh and blood we speed past every day.

A CHRISTIAN PERSPECTIVE

Driving a vehicle alongside oncoming traffic is always dangerous. We command a three-thousand-pound block of steel and glass (or a forty-five-hundred-pound SUV) at high speeds, often with little separation other than a painted line on the road. Split-second miscues accelerate quickly into irreversible tragedies and lifelong, haunting regrets. The tools we use in our lives put others in the way of harm, and one little slip can change lives forever.[4] Texting while driving and living the rest of our lives with the blood of innocents on our hands are more closely related than we like to think.

What laws cannot stop, Scripture addresses as matters of the heart. Jesus boiled down the Christian life to two basic questions: "How do I love God?" and "How do I love my neighbor?"[5] And when Jesus was asked to define "neighbor," he pointed to a road.[6] In the

---

4. Deut. 19:4–10.
5. See Matt. 22:37–40.
6. Luke 10:29–37.

digital age (as was true in the predigital age), remote people and concerns can command our undue attention, blinding us to the immediate needs around us. As we drive, our phones *ping*, our brains get a shot of dopamine, and very often our decisions express our own neighbor negligence. We assume we can ignore the people we see in order to care for the people we don't see, but that idea is all twisted backward.[7]

We sin with our phones when we ignore our street neighbors, the strangers who share with us the same track of pavement.

VIRAL ANGER

Texting and driving is one example of the main point of this chapter. We are quick to believe the lie that we can simultaneously live a divided existence, engaging our phones while neglecting others.

A second example of this fracturing is our online conflict.

Our bodies distinguish us from one another and mark off our existence in the world. In the digital realm, we lose this key reference point.[8] We lose sight of one another, and when we do, anger boils more quickly.

We are more likely to bubble with rage toward others screen to screen instead of face to face, and researchers call this phenomenon "anonymous anger." The steam of anger finds quick release in words thumbed into our phones. It is too convenient to vent our rage in public now. On top of this, there are three other culprits: "relative anonymity, a lack of authority and consequences, and solipsistic introjection—the theory that, subconsciously, talking on a computer can seem more like we're talking to ourselves than to real people." In other words, "It's very difficult to link words on a screen with the reality that there's a living, breathing human on the other end of the connection."[9] Online anger is a consequence of the division in

---

7. 1 John 4:20.

8. Alastair Roberts, "Twitter Is Like Elizabeth Bennet's Meryton," *Mere Orthodoxy*, mereorthodoxy.com (Aug. 18, 2015).

9. Nick English, "Anger Is the Internet's Most Powerful Emotion," Greatist, greatist.com (Sept. 18, 2013).

our lives—our attention is divided, our minds are divided, and our digital personas are separated from our flesh and blood.

These divisions lead to avoidable misunderstandings and short fuses online. Our typing thumbs lack empathy without living faces in front of us. It is much easier to slander an online avatar than a real-life brother.

But online anger is not merely pervasive; it's also contagious. I've been immersed in the world of social media long enough to discover that the single most important determining factor about whether what I publish online will get hot, spread virally, and reach new pockets of readers is my success at igniting a heated debate. Studies back this up on a more personal level, showing that a joyful comment is likely to bless your following but not go much further, whereas a furious comment is far likelier to spread outside of your following and enrage many more people. "Anger is a high-arousal emotion, which drives people to take action," said one researcher of this trend. "It makes you feel fired up, which makes you more likely to pass things on."[10]

Rage spreads.

THE JOY OF FELLOWSHIP

If anger is the viral emotion of online disembodiment, then *joy* is the Christian emotion of embodied fellowship, and two apostles prove it: John and Paul. John closed one of his ancient handwritten letters with a line of enduring relevance for those of us who now write with our thumbs: "Though I have much to write to you, I would rather not use paper and ink [modern technology for John]. Instead I hope to come to you and talk face to face, so that our joy may be complete" (2 John 12). John used technology to communicate, but he knew that his letter was only part of the communication. It was a way of expressing anticipation; face-to-face fellowship had to follow. Paul makes the same point in two of his letters.[11]

---

10. Matthew Shaer, "What Emotion Goes Viral the Fastest?" *Smithsonian* (April 2014).
11. Rom. 15:32; 2 Tim. 1:4. This is rooted in the eschatological hope of 1 Thess. 2:19–20.

So why do two apostles tell us their joy is bound up with embodied fellowship? "I think it has to do with the engaging of personalities," Douglas Groothuis, a professor of philosophy at Denver Seminary, told me. "Our personality will come through to some extent in an email message or a tweet. But we are holistic beings: we have feelings, thoughts, imaginations, and bodies." When we remove part of our embodied personhood, misunderstandings become easier. When we trade our physical arms that cross, eyes that linger, ears that detect sarcasm, and vocal tones that imply patience for the two-dimensional avatar, we invite misunderstanding and tension. "So I think the 'fullness of joy' comes with one personality interacting with other personalities in terms of voice, touch, appearance, and timing. Sometimes it is time just to be quiet with people, or to cry with people, or to laugh with people."[12]

On top of this, eye contact is one of the most powerful forms of social bonding possible, forging trust between people in a complex phenomenon whereby people can sync their minds and gain mutual understanding, learning, and sharpening in ways impossible through digital devices.

There are certainly many other reasons to cherish face-to-face meetings, but these passages from the apostles leave us with an important point we need to remember in our digital communications technology. All writing that is remote—like the ancient letter, the modern text message, or this book—is more like ghost-to-ghost communication than person-to-person interaction. Yes, there is *something* of us in written words, but not *everything* in true fellowship can be typed out on phone screens and sent at the speed of light through fiber-optic cables. This is the reality of communication. Joy is a precious emotion of our integrated existence. Joy brings our attention, our minds, and our flesh and blood together into face-to-face fellowship—eyeball-to-eyeball love. The Christian's challenge is to love not in tweets and texts only, but even more in deeds and physical presence.[13]

---

12. Douglas Groothuis, interview with the author via phone (July 3, 2014).
13. 1 John 3:18.

COMPOUNDED EMBODIMENT

In the smartphone age, when our cognitive actions are separated from our bodily presence, we tend to overprioritize the relatively easy interactions in the disembodied online world and undervalue the embodied nature of the Christian faith.

From the opening narrative of God becoming flesh, the New Testament is thick with the idea of embodiment. Keep reading, and Scripture describes the nature of God's people: we are individual members of the church, and our unity amid diversity finds expression in metaphors of the multisensory and multifunctional nature of the human body.[14] Keep reading, and Paul encourages holy kisses (awkward!).[15] He also warns us not to neglect our gathering together,[16] and focuses on two common church celebrations: baptism and the Lord's Supper. Both sacraments are essential to our gatherings and contain multiple layers of compound embodiments. We cannot be baptized or feast at the Lord's Table on our phones.

It is an act of obedience for a follower of Christ to be immersed under water. For me, it happened in a temporary hot tub set up on a church stage in the middle of winter, when my death to sin and new life in Christ were reenacted. At one level, it was purely metaphorical: as I was pushed under the water, my union with the physical death of Christ was symbolized. As I surfaced, my spiritual resurrection in the physical resurrection of Christ was depicted. The *spiritual* meaning of my water baptism was not possible without the *physical* death and *physical* resurrection of Christ. But the drenching of my baptism did not merely symbolize a past or present spiritual reality in me. I am now certain that when my *physical* death arrives and my body is placed in the ground, it will be planted like a seed, waiting to spring eternally in physical resurrection. The metaphorical act of my baptism symbolized what is possible only by the physical reality

---

14. 1 Cor. 12:12–31.
15. Rom. 16:16; 1 Cor. 16:20; 2 Cor. 13:12; 1 Thess. 5:26; 1 Pet. 5:14.
16. Heb. 10:24–25.

of Christ, and my spiritual union with him guarantees my physical future.[17]

The Lord's Supper is another practice for the gathered church, assembled in physical unity, not given over to interpersonal factions. In this unity, we imitate Christ. On the night of his arrest, Jesus tore the bread and poured the cup, and said it was his body broken and blood shed for sinners. Every time we reenact Jesus's pattern, we remember Christ (now unseen) and proclaim his death until he returns (then seen)—affirming that he is as real as the cup and bread in our hands. And if any one of us should approach this table selfishly or unworthily, we risk physical sickness, and even bodily death, as a consequence![18]

In our bodies, we carry around the death of Jesus, so we can lay down our lives for our brothers and sisters in Christ.[19] Every unseen spiritual reality in the Christian life, and every physical practice in the church, is rooted in the physical realities of our Savior—that he was and is God incarnate. He lived, he walked, he ministered, he was crucified, he died, he was buried, he was raised to new life, he is now seated in heaven, and he is soon to return. If these physical realities are mere fiction, then our hope and faith—from head to toe—are entirely futile.[20]

The modern-day mantra we hear so often—"I will follow Christ, but don't bother me with organized religion"—is symptomatic of the disembodied assumptions of the digital age. In reality, the Christian life could not be more embodied. To ignore all these facts, and to prioritize our disembodied existence online, is nothing short of "conniving at dehumanization."[21]

MUDDY PIXELS

The implications of our lives in these bodies will be considered again later in the book. For now, it is enough to go back to the point where

---

17. Rom. 6:1–11.

18. 1 Cor. 11:17–34.

19. 2 Cor. 4:10–11; 1 John 3:16.

20. 1 Cor. 15:14.

21. Medri Kinnon Productions, "N. T. Wright on Blogging and Social Media," Vimeo, vimeo.com (July 20, 2009).

we started: the epidemic of texting and driving (among many other epidemics) is an attempted escape from the limits of our flesh-and-blood nature. We try to break through the boundaries of time and space, and we end up ignoring the flesh and blood around us.

In reality, we are finite. We assume that we can drive cars and read and write on our phones all at the same time, but we are weaker than our assumptions. To exist is to be walled in by physical limitations—boundaries and thresholds that limit what we can perceive and accomplish. When we always see our lives through glass, we forget that we are made of flesh and blood.

In truth, we are finite flesh and blood living among finite flesh and blood. And if studies are right, large numbers of smartphones have trace amounts of fecal matter on them. I read the news reports and chuckle at the grossed-out comments that follow. We are creatures made from mud, holding pieces of glossy glass and trying to preserve their shimmering cleanliness with state-of-the-art cases and microfiber cloths. This is impossible. We are not technology. We are not smooth, clean, and indestructible like man-made crystal. No. We are easily scratched. We are born broken. We are dust and water, chemicals and germs, and everywhere we go we leave oily blots on everything we touch. It is almost impossible to miss the juxtaposed parody between our dusty selves and glistening pixels. We smudge technology because we are not machines. We are creatures made in the image of the supreme Creator, and we are made to share embodied joy together, in his name.

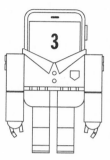

WE CRAVE IMMEDIATE APPROVAL

In the digital age, we can ignore bodies, but we can also abuse them.

Meet Essena O'Neill, who, as a nineteen-year-old Australian model, accumulated five hundred thousand Instagram followers. Once poised to make a career from online endorsement deals, in 2015 she called it quits, deleted most of her pictures, and revised the remaining descriptions to unmask the true motives behind the images (mostly sponsored product placements). Why the drastic move?[1] Essena had come to see that her online life was hollow, fake, and self-centered.

"Over-sexualization, perfect food photos, perfect travel vlogs—it is textbook how I got famous," she admitted.[2] But it was all part of a downward spiral she came to regret. "Everyone goes through life differently, myself growing up with social comparing so easily available. It consumed me. . . . I spent [ages] 12–16 wishing I was someone else. Then I spent [ages] 16–19 constantly molding myself, editing and self-promoting the 'best parts of my life'—which turned into a massive career based on numbers and how I looked aesthetically."[3]

---

1. Some critics say it was a publicity ploy for attention. In this project, I trust her stated intentions.

2. Megan McCluskey, "Instagram Star Essena O'Neill Breaks Her Silence on Quitting Social Media," *Time* magazine (Jan. 5, 2016).

3. Essena O'Neill, "Dear 12 Year Old Self (re-upload)," YouTube, youtube.com (Nov. 8, 2015).

Today, Essena said, "I simply no longer want to compare my life with anyone else's edited highlights. I want to put all of those hours I looked into a screen into my real life goals, personal relationships, and aspirations. I'm over this celebrity culture and obsession. It's silly, and for the most part, internally lonely and fake."[4]

Most tragically, Essena admitted that she had mindlessly offered her body up for public admiration, posting selfies in order to be told she was beautiful and attractive. "Being born into this screen-dominated age, we are taught to mold ourselves in order to gain the most social validation [likes, views, and followers across social media]," she said. "I've simply taken myself out of the sculpting studio. I don't want to look to others for how I should live, speak, and create."[5]

In the end, she said, "I was a living paradox of conditional self-love and constant self-hate. Basically, my self worth relied on social approval." She assumed that she could satisfy her heart by becoming "Facebook famous" or "Instagram famous," but as her fame grew, her life felt more and more shallow and contrived. The popularity made her feel—in her words—trapped in a cycle that became more empty, lonely, hateful, jealous, and insecure.[6] And nothing traps people in unhealthy social-media patterns like personal insecurity.[7]

She's not alone. Meet "Jasmine," a twenty-something woman aspiring to Instagram fame, who spoke out, but only under an alias because she was still in the game and was too embarrassed to admit it. The identity she projected was costly, and she found herself drowning in credit card debt. "I buy a lot of things to maintain my image," she said. "I pay for meals out, new bikinis (I've never photographed

---

4. Ibid.

5. Essena O'Neill, "Social Media Addiction and Celebrity Culture," letsbegamechangers.com (Oct. 30, 2015). This and the following quotations from Essena O'Neill appeared in materials on her website, letsbegamechangers.com, at the time of writing. Prior to publication, that site was taken down. Interested readers can find the quotations by searching letsbegamechangers.com through web.archive.org.

6. Essena O'Neill, "Liked," letsbegamechangers.com (undated).

7. "The students I interviewed who suffer from insecurity, who have anxiety about their social standing, who fret about how they are seen by others, are the ones who are drowning on social media." Donna Freitas, *The Happiness Effect: How Social Media Is Driving a Generation to Appear Perfect at Any Cost* (New York: Oxford University Press, 2017), 20.

the same one twice), beautiful printed dresses nearly once a week, fresh flowers religiously once a week, etc. . . . I spend money to make my life look a certain way, and I get a rush from looking that way, but my credit cards do not share my enthusiasm." Her $3,400 credit card debt was mounting. She couldn't pay it off, but she couldn't stop the compulsive buying. "As I'm writing this, I'm eating the sushi I bought on my way home, photographed fifty times, posted, and got 231 likes on so far. I plan on telling my parents about this when I go home next weekend so they can yell at me and force me to stop, because I know they'll absolutely freak out. I know exactly how stupid what I'm doing is, but I just need someone to tell me, I guess."[8]

Essena and Jasmine are extreme examples of the smartphone temptations we all face every day. Although we might not have half a million followers or mounting credit card debt, we can obsess over our image management just as much, and just as easily slip into behaviors that we hardly notice until we're in too deep.

HERO VERSUS CELEBRITY

Essena, Jasmine, and every other Instagram or YouTube celebrity is a modern icon of what historian Daniel Boorstin warned us about fifty-six years ago. He predicted that after the arrival of "the Graphic Revolution," which exploded the ability to mass produce and edit images of people in film and in print (and now online), our *heroes* would be replaced by *celebrities*.[9] He was right.

Heroes are men and women of character, known for acts of valor and celebrated long after their deaths. Time, not image, makes heroes. Heroism goes mostly unseen in the moment, and our heroes, at least in the case of our presidents, appear nearly lifeless on our currency, intentionally washed of glamour. Every

---

8. Jasmine, "The Financial Confessions: 'My "Perfect" Life on Social Media Is Putting Me in Debt,'" *The Financial Diet*, thefinancialdiet.com (April 12, 2015). But here's the catch: scaling up online fame may not alleviate the problem, but only make finances harder. In the words of one writer: "Many famous social media stars are too visible to have 'real' jobs, but too broke not to." Gaby Dunn, "Get Rich or Die Vlogging: The Sad Economics of Internet Fame," *Fusion*, fusion.net (Dec. 14, 2015).

9. Daniel J. Boorstin, *The Image: A Guide to Pseudo-Events in America* (1961; repr., New York: Vintage, 1992), 45–76.

culture has its heroes, because we want to know that humanity is potentially great. So we immortalize our minted heroes on our bills, coins, and stamps.

But we lost our patience in waiting for new heroes just as the Graphic Revolution arrived, so we coined new icons. The dominance of images in the media (and now a hyperabundance of digital images) meant that waves of celebrities could be created, rejected, and replaced. We turned to celebrities who were "fabricated on purpose to satisfy our exaggerated expectations of human greatness." Unlike the hero, the celebrity is newsworthy simply for his visible charm, his spectacle of glamour, writes Boorstin. In fact, "anyone can become a celebrity if only he can get into the news and stay there." It is all about time, and that is the greatest contrast of all. "The passage of time, which creates and establishes the hero, destroys the celebrity. One is made, the other unmade, by repetition."[10]

WARHOL'S IMAGE FACTORY

Perhaps no artist exploited this image-driven phenomenon more than Andy Warhol (1928–1987), who devoted his life to replicating powerful images in pop art. He was a product of the Graphic Revolution and one of its masters. Long before the smartphone made it technologically convenient (or socially normal), he carried around sound recorders and Polaroid cameras in public as a sort of buffer between himself and the world. When he turned the Polaroid on himself, he essentially invented the selfie.

"If there is a current animating Warhol's work, it is not sexual desire, not eros as we generally understand it, but rather desire for attention: the driving force of the modern age," writes Olivia Laing. "What Warhol was looking at, what he was reproducing in paintings and sculptures and films and photographs, was simply whatever everyone else was looking at, be it celebrities or cans of soup or photographs of disasters, of people crushed beneath cars and flung into

---

10. Ibid.

trees." By reproducing eye-grabbing images, he was tapping human attention, and this drew him to the ultimate image-reproducing medium of his day, television. Warhol believed if he could get into television and replicate himself in every living room in flickering images, he would feel accepted. "That's the dream of replication," Laing says with stinging insight, "infinite attention, infinite regard."[11] But it's a lie of the celebrity culture: replicated images of the self will never deliver the intimacy they promise.[12]

Warhol's image replication foreshadowed a moment when all of us could easily duplicate images of ourselves digitally through our phones in selfies and in avatars that reappear every time we speak online, making it possible for each of us to grab attention and taste fame, even for a flickering blink.

And yet, writes Laing, all these attempts at repetition and fame really become "intimacy's surrogate, its addictive supplanter."[13] The digital world through our phones allows us the tools of self-replication and the hope that we can garner infinite attention and infinite regard from others, and, in that way, achieve a sort of online fame. But online attention proves to be an incapable substitute for true intimacy, and the addiction to a crafted online image renders true intimacy impossible.

THE COMFORTS OF ONLINE FRIENDS

Many Christians avoid these vanities and prove themselves skilled at using social media and their phones to build relationships with people they know face to face. For them, Sundays are a sweet time to connect in person. But some of us use our phones more often to connect with people we don't regularly see in person; we might show up on Sunday and feel out of place among strangers. Have you ever wondered why it feels so natural to communicate with others

---

11. Olivia Laing, *The Lonely City: Adventures in the Art of Being Alone* (New York: Picador, 2016), 245.

12. See Tony Reinke, "Selfies and Polaroids of Intimacy: Andy Warhol and My Smartphone," Desiring God, desiringGod.org (April 7, 2016).

13. Laing, *The Lonely City*, 243–44.

online but it sometimes feels awkward to communicate with others at church on Sunday mornings? Many factors are at work.

First, in the online world, we can break free from our physical limitations (if we want to). We can present ourselves as older, or younger, than we really are. We can monetize our bodies as a means of grabbing attention and selling products online—if we have the right physique. If not, if we are overweight or unsightly, our bodies can be shielded from digital eyes. If we are physically disabled, we can hide our wheelchairs completely from our digital friends. The physical defects, limitations, and awkwardnesses that we are born with, or that we now live with, can all be dissolved and glossed over online. Hiding our unflattering features is very natural and easy online, but excruciatingly hard and unnatural offline, in healthy local churches and honest friendships. Self-editing is less possible in genuine face-to-face relationships. There is no Valencia filter for the real-life you. Without honestly acknowledging these online tendencies, we will continue to think local-church awkwardness is a strange feeling to be resisted rather than a precious means to reshape us.

Second, in the online world, we can separate ourselves from people who don't think like us and gravitate toward people who do. This is one reason why I love to write online. Reading and writing in the instantaneous digital world of online social networks is a means to profound Christian fellowship. We can disclose things very near to our hearts and our core fears and convictions, and some of our closest friendships can be forged and maintained on our phones with people around the world. But as mentioned in the last chapter, there can be a serious downside to online-only fellowship.

My research on this point eventually brought me to northern England, to Alastair Roberts, a studious thirty-six-year-old theologian and eloquent writer laboring in the fields of biblical theology and contemporary ethical issues, including our relationship to developing technologies. Roberts is also a longtime blogger who wisely warns of one toxic danger that threatens our online communities:

The Internet can enable us to form connections with people with whom we have extremely particular things in common, making possible highly stimulating, enriching, and deepening interactions. I wouldn't be where or who I am today were it not for online interactions, sustaining and helping me to develop a perspective that often bears little relation to my immediate contexts over the years.

This said, while I have undoubtedly gained an immense amount from these, I have frequently found them to be a retreat from the challenge of actual relationships with Christian neighbors with whom I differ, a temptation amplified for me by virtue of the fact that I can naturally be an extreme introvert, prone to reclusiveness. When you know that there is a place where everyone largely agrees with and values you, you can develop a reluctance to go to a church where you are not so valued, understood, or appreciated. The narcissism that can be characteristic of romantic ideals—romantic ideals that can actually drive us away from our real partners into escapist and emotionally comforting reveries—can also cause us to replace the concrete relationships of our given contexts with idealized communities in which we can forgo the struggles associated with the transformation of actual communities and the need to adapt to and be vulnerable to others.[14]

We easily settle into digital villages of friends who think just like us and escape from people who are unlike us. Our phones buffer us from diversity, warns Roberts. Although "generational differences are fundamentally constitutive differences for the human race . . . new media is one of many ways our elders are rendered invisible."[15] And it's not only our elders, but also the impoverished, the cognitively disabled, children, the less educated, the less literate, the less cosmopolitan, and non-Westerners. In effect, our online communities "render invisible the majority of the human race."[16]

---

14. Alastair Roberts, interview with the author via email (Jan. 23, 2016).
15. Alastair Roberts (@zugzwanged), Twitter, twitter.com (Jan. 18, 2016).
16. Roberts, interview with the author via email (Jan. 23, 2016).

In fact, our online communities of like-minded friends are often marked by a "positive feedback loop," where "affirmation and assent merely reinforce existing prejudices. In such contexts, communities become insular, echo chambers of accepted opinion, closed to opposing voices," which means they breed a "homeostatic stifling of difference."[17] Communities that fail to embrace the benefits of disagreements and fail to work through tensions and differences tend to become homogeneous and unhealthy, because they "tend to have exaggerated blindspots and unaddressed weaknesses."[18]

But perhaps we can press in further. Just as it's hard to grow together as a team when each player is preoccupied with individual performance and popularity,[19] so too it's hard to grow as a family when children bring the hyperapproval climate of school into the home through their ubiquitous phones.[20] Boring team meetings and boring family times are truly opportunities for personal growth in places of unconditional love, providing the soul a respite from the now unceasing demands of social approval.

Maybe this is a key function of church attendance in the digital age. We must withdraw from our online worlds to gather as a body in our local churches. We gather to be seen, to feel awkward, and perhaps to feel a little unheard and underappreciated, all on purpose. In obedience to the biblical command not to forsake meeting together,[21] we each come as one small piece, one individual member, one body part, in order to find purpose, life, and value in union with the rest of the living body of Christ.

This feeling of awkwardness, this leaving the safety of our online friendships, this mingling with people we don't know or understand

---

17. Alastair Roberts, "Twitter Is Like Elizabeth Bennet's Meryton," *Mere Orthodoxy*, mereorthodoxy.com (Aug. 18, 2015).

18. Roberts, interview with the author via email (Jan. 23, 2016).

19. NFL head coach Sean Payton, when asked about his greatest coaching challenges, pointed to social media and fantasy football; the one isolates individual player performances and the other generates unending hype for over- and underperforming players. In team meetings, players itch to get back to their phones. See "Sean Payton: That's the Biggest Challenge as a Coach in Today's Game . . . ," Coaching Search, coachingsearch.com (Feb. 21, 2016).

20. See Suzanne Franks, "Life Before and After Facebook," *The Guardian* (Jan. 3, 2015).

21. Heb. 10:24–25.

in our local churches is incredibly valuable for our souls. Church is a place for real encounters with others and for true self-disclosure among other sinners. In the healthy local church, I do not fear rejection. In the healthy local church, I can pursue a spiritual depth that requires agitation, frustration, and the discomfort of being with people who conform not to "my" kingdom but to God's. The challenge for us is to "cherish corporate worship, that most counter-cultural of practices, for which no virtual substitute can be found."[22]

GLORY VERSUS APPROVAL

This discussion raises the question of approval. The online celebrity culture is driven by glory, praise, and approval, but so is the Bible. God's story is loaded with awe, admiration, and wonder. The tug and pull of existence is tethered to the powers and pressures of glory. We live in a story of competing pleasures and displeasures, between the joy of approval and the depression of disapproval. So as Christians, how do we make sense of these tensions in the digital age?

Christ helps us discern between the glory of man and the glory of God in a story told in John 12:27–43. It was a spoiler-alert moment: Jesus had entered Jerusalem as a King on a donkey after a Jewish feast had drawn large crowds to the city. To quiet and summon everyone, God spoke from the heavens. Then Jesus stood to foretell the pinnacle of history, which was about to unfold: he would be raised up on a cross in death and later raised up in the resurrection. On the soon-approaching weekend, the Creator's entire timeline for the universe would hinge and take a cosmic turn.[23] Those who would understand and lean toward belief in Christ would walk in the light. Those who would disbelieve would continue to live in darkness. And darkness would dominate.

Christ is the revelation of God's glory, the image of the invisible God, but the majority of the religious leaders failed to see him for who he was. The few who did were plagued by a weak faith. No matter

---

22. Oliver O'Donovan, interview with the author via email (Feb. 10, 2016).
23. Heb. 1:2; 9:26.

how many miracles Christ pulled off (even raising the dead), most of the leaders flatly refused to celebrate the Messiah. Why? What could possibly mute mouths on the brink of the climactic moment in cosmic history?

"Many even of the authorities believed in him," John tells us, "but for fear of the Pharisees they did not confess it, so that they would not be put out of the synagogue; for they loved the glory that comes from man more than the glory that comes from God" (John 12:42–43).

Why was it so hard for them to celebrate Christ? Simple—public approval forbade it. If you follow Christ, the world will unfollow you. You will be shunned. You will be despised. If the glory of man is your god, you will not celebrate the glory of Christ. Or, if you come to Christ and treasure his glory above all other glory, you will be forced to forfeit the buzz of human approval. Christians today still face real-life glory wars and real-life tensions inside the digital world. So what do we fear more, the disapproval of God or the disappearance of our online followers?

TRUE APPROVAL

The approval and affirmation we seek online is absurd because it misunderstands how approval works in God's economy.

First, the itch for human approval ultimately renders faith pointless.[24] Why? Because faith is the act of being satisfied with Christ, says John Piper, "and if you are bent on getting your satisfaction from scratching the itch of self-regard, people's affirmation, you will turn away from Jesus, because you can't serve two masters." In other words, he says, "In a solid, God-chosen relationship with Jesus, man's disapproval cannot hurt you and man's approval cannot satisfy you. Therefore, to fear the one and crave the other is sheer folly."[25] It is unbelief.

Second, the test of authenticity for our lives is not determined by

---

24. John 5:41–45.
25. John Piper, interview with the author via Skype, published as "Gospel Wisdom for Approval Junkies," Desiring God, desiringGod.org (March 15, 2016).

the applause of man, but by the approval of God.[26] We cannot commend ourselves. God commends us.[27] He searches us. He knows our every motive, even our motives for ministry.[28]

The sad truth is that many of us are addicted to our phones because we crave immediate approval and affirmation. The fear we feel in our hearts when we are engaged online is the impulse that drives our "highly selective self-representation."[29] We want to be loved and accepted by others, so we wash away our scars and defects. When we put this scrubbed-down representation of ourselves online, we tabulate the human approval in a commodity index of likes and shares. We post an image, then watch the immediate response. We refresh. We watch the stats climb—or stall. We gauge the immediate responses from friends, family members, and strangers. Did what we posted gain the immediate approval of others? We know within minutes. Even the promise of religious approval and the affirmation of other Christians is a gravitational pull that draws us toward our phones.

THE COST OF SEEKING APPROVAL

This approval addiction must be why Jesus expressly warns us not to seek human praise by our obedience. He warns us not to flaunt our works online in order to be praised by others: "Beware of practicing your righteousness before other people in order to be seen by them, for then you will have no reward from your Father who is in heaven" (Matt. 6:1).

Consider one example. Imagine setting aside a few weeks of your summer vacation to travel on dirt roads and bump around in loud jeeps, winding deep into remote jungle villages in Central America. You risk fevers, diseases, and heatstroke, all in order to help build an orphanage for twenty destitute kids. At the end of the month, you step back, take a selfie with your handiwork in the background, and post it with pride on Facebook. Poof!—the reward is gone. Think

---

26. Rom. 2:29.
27. 2 Cor. 10:18.
28. 1 Thess. 2:3–5.
29. Roberts, interview with the author via email (Jan. 23, 2016).

about it. In one humble-brag selfie, the trade is made—eternal reward from God is sold for the porridge of maybe eighty likes and twelve comments of praise. (Context is not the point; we do this same sort of thing with pictures of an open Bible in a coffee shop.)

Could it be that my application of Jesus's words is too rigid and not focused on the heart intent of the act? Perhaps, but shouldn't we check ourselves through concrete examples like these? We must agree that at some level, Jesus said that publishing our good works online for our followers to see is all the reward we'll get.

The trade is horrible. "You lose something great, and you gain something pitiful," Piper explains. "What do you gain? You gain the praise of man. You want it? You get it. It's like a drug. It gives a buzz, and then it is gone. You have got to have another fix. And it leaves you always insecure. You are always needy of other people's praise in order to be happy or to feel secure. You are never satisfied."[30] We wake up each day hungrier than ever for validation.

The buzz of social approval has conditioned us to feed on "regular micro-bursts of validation given by every like, favorite, retweet, or link."[31] This new physiological conditioning means that our lives become more dependent on the moment-by-moment approval of others. The problem is not just that we need to turn away from these micro-bursts of approval, but that we must deprogram ourselves from this online hunger.

If we don't detox these habits, we will go on seeking intimacy by reproducing ourselves, bingeing on man's approval, and starting each day with an approval hangover. Then we need the antidote of new affirmation from our friends to keep convincing ourselves that our lives are meaningful. This is tragic. This is wasted reward. The solid praise we expect from God is based on actions now largely unseen; the whimsical praise we seek online is based on what we project.[32] We cannot neglect this contrast.

---

30. John Piper, interview with the author via Skype, published as "Incentives to Kill My Love of Human Praise," Desiring God, desiringGod.org (Aug. 25, 2014).

31. Alastair Roberts, email to the author (Feb. 22, 2016). Shared with permission.

32. Rom. 2:28–29.

DON'T WASTE YOUR APPROVAL

Smartphones prick the primitive human impulse for appreciation—self-replication in order to be seen, known, and loved—through constant contact with other seekers of affirmation. This is one reason why we find it so hard to put our phones away. We fear one another, and we want admiration from one another, so we cultivate an inordinate desire for human approval through our social media platforms. For those of us who struggle here, Jesus's warning is very clear: "Whoever loves [his social network] more than me is not worthy of me" (Matt. 10:37).

He can say such a hard truth because the truly heart-satisfying affection we need is ultimately in God, in beautiful promises such as those in Psalm 139, where our souls are soaked in layers of precious truths about God's acceptance and love for us. His power washes over our lives, and his presence overpowers every small gain of digital notice and acceptance we seek online. He reminds us that our lives are not sustained by the fickle approval of others to our self-replication; they are sustained by God's sovereign expanse over all things.

We cannot continually chase the lure of public praise and affirmation by self-replication. Such a desire will kill us spiritually, and Paul signaled why. In God's economy, approval is something we must wait for. Those who feed on little nibbles of immediate approval from man will eternally starve. But those who aim their entire lives toward the glory and approval of God will find, in Christ, eternal approval.[33]

The stakes are that high.

If you want to become an "Instagram celebrity," if you crave fame and seek it through self-promotion, I plead with you to stop. The urgency that you feel and that drives you online is caused by your fear of being unreplicated, unseen, unloved. Each day you feel as if you are losing your grip on your online status unless you deliver crowd-pleasing content. Stop trying to impress the online world with your body or your wit, all for the sake of a few likes of affirmation. Vain

33. Rom. 2:6–11.

glory will not satisfy your heart; it will only intensify your craving for human praise.

Daniel Boorstin was right all along: we must reckon with time. Is your heart set on becoming a celebrity in this life or a hero in the next? Is time your daily nuisance, threatening to erode your significance, or is it your friend? Do you want your approval and fame now, or can you wait for an eternal crown? We all must answer these questions, and how we answer them will determine whether our souls find health in Christ or sickness in the spotlight.

As we fight against the lure of self-glorification, Jesus, Paul, and Peter all plead with us: Don't waste your approval. Don't crave the approval of man online. Don't flaunt your righteous deeds in the cyber world. If we miss their warnings, we will make a cosmically foolish mistake, with eternal regret to follow.

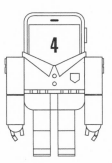

WE LOSE OUR LITERACY

Our entire faith is built on a book, and inside that book are sixty-six smaller books. Our spiritual life is fed by books within books, like Ezekiel's wheels inside wheels. And new Christian books are released every day around the world. Books are a big deal for Christians. We treasure the press. Publishing is part of gospel mission. Wherever the gospel has spread, so has literacy.[1]

Yet in the digital age, books have become more vulnerable to the label *boring*. Compared to the latest game or streaming television series, staring at black and white shapes (like these) for several hours seems like a silly investment. We have been initiated into a kind of entertainment-convenience that makes books feel downright outdated, inconvenient, and far too demanding.

The statistics show that Christians who struggle to read books are struggling to break free from poor smartphone habits as one root cause.

LOSING OUR LITERACY?

In my survey of eight thousand Christians, which I mentioned in chapter 1, I asked: How many nonfiction books (of at least one

---

1. In this chapter, I argue for the value of book reading in the Christian life, but I do it rather quickly. A fuller bibliophilism can be found in my book *Lit! A Christian Guide to Reading Books* (Wheaton, IL: Crossway, 2011).

hundred and fifty pages) have you read in the past twelve months?[2] As expected, the results for Christians were a little above the national average:

	Men	Women
0–2 books	41%	47%
3–6 books	33%	34%
7+ books	26%	20%

Next, I asked: In general, do your smartphone and current use of social media cause you to read more books or fewer books, or not cause any noticeable difference in the number of books you read?

I discovered two interesting facts. First, a fairly large number of Christian smartphone users are becoming *more voracious* readers of nonfiction books. In this case, social media and our online communities are powerful forces to encourage Christian literacy, a phenomena I understand firsthand. I read more books now than ever in my life because social media connects me with discerning readers who also love to read and share great books.

Second, I made a less encouraging discovery. Far more commonly, I heard that smartphone users are reading fewer books. A large portion of my sampling (about three thousand out of eight thousand respondents) said their use of their phones negatively impacts the number of books they read. Figure 2 on page 81 summarizes the survey results, with the percentages of those who said their smartphone use has caused them to read fewer books (black) or more books (gray) broken down by age/gender demographics.

The missing middle between these two polar responses is occupied by those who sense no connection between their book reading and their smartphone use (which comprised about 50 percent of both men and women).

---

2. A nonscientific survey of desiringGod.org readers, conducted online via social-media channels (April 2015).

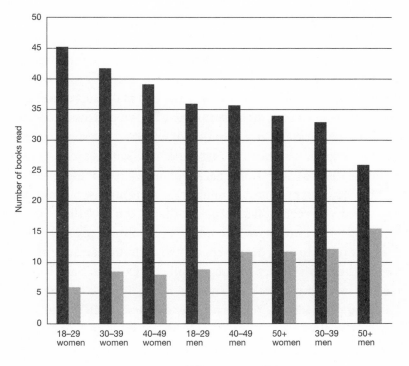

Figure 2. Christians and reading

Still, the fact that three thousand respondents said they now read fewer books as a result of their phones and social media (with women accounting for 56.8 percent and men 43.2 percent) shows that as phones permeate our lives, it is becoming increasingly difficult for a substantial percentage of young Christians to read books.

THE ENDLESS COCKTAIL PARTY

It is a matter of attention, and in the digital age, our attention is a commodity worth money. Let me illustrate this with an offline situation. I live close to one of the largest malls in the world, and my enjoyment of walking around that mall would be improved without the kiosk vendors. Few people go to malls to browse the kiosks. (I sure don't!) And the vendors know it. Their only hope for making a sale is to grab your attention. They cannot sell you anything unless they can distract you, and they cannot distract you unless you make eye contact. (That means that the key is to avoid eye contact.) Malls are a metaphor for our phones: they are bustling commercial centers for the street performers of digital allurements. In their own ways, all of our social media compete for more and more of our attention, at the cost of the sustained focus we need to read books. This is because texts, snaps, and tweets are part of an endless cocktail party of multiple conversations, suggests *New York Times* columnist David Brooks. And have you ever tried to read in the middle of a party? And what about in a party that never ends?

"The slowness of solitary reading or thinking means you are not as concerned with each individual piece of data," Brooks writes. "You're more concerned with how different pieces of data fit together. How does *this* relate to *that*? You're concerned with the narrative shape, the synthesizing theory, or the overall context. You have time to see how one thing layers onto another, producing mixed emotions, ironies and paradoxes. You have time to lose yourself in another's complex environment." Brooks calls this discipline "crystallized intelligence"—"the ability to use experience, knowledge, and the products of lifelong education that have been stored in long-term memory. It is the ability to make analogies and comparisons about things you have studied before. Crystallized intelligence accumulates over the years and leads ultimately to understanding and wisdom."[3]

Such a skill requires separation from the digital cocktail party so we can activate our sustained linear attention and engage our minds. The fragmentary nature of the online world makes this type of concentration difficult to maintain—all by design.

With so much at stake, corporations are refining the art of attention capturing with a growing field of technological expertise called "captology," a nickname for "computers as persuasive technology." Captologists study ways of using smartphones to capture attention and to adjust behavior patterns.

One trick works like this: the more I *like* and click online, the more precisely web algorithms feed me images, ideas, and products tai-

---

3. David Brooks, "Building Attention Span," *The New York Times* (July 10, 2015), emphases added.

lored to my previous engagement. It may seem I am simply stumbling over a litany of randomly scattered things online, but what's offered up to my eyes today is increasingly aligned to the bread-crumb trail I left behind in my digital diet yesterday (for good or ill). So what I see now has been tailored to what I *liked* in the past, creating a custom-built vortex of content, a swirl of new objects, filling my screen as I flick and scroll, all with the aim of keeping my eyes glued to the screen by feeding very specific appetite patterns of my craving heart and ultimately reinforcing my smartphone obsession.

"The media have become masters at packaging stimuli in ways that our brains find irresistible, just as food engineers have become expert in creating 'hyperpalatable' foods by manipulating levels of sugar, fat, and salt," writes Matthew Crawford, a writer and senior fellow at the Institute for Advanced Studies in Culture at the University of Virginia. "Distractibility might be regarded as the mental equivalent of obesity." Without the ability to focus our minds, our attention is led by others, and we are easily captured by "the omnipresent purveyors of marshmallows"—the alluring distractions on our phones. Crawford asks, "What sort of outlier would you have to be, what sort of freak of self-control, to resist those well-engineered cultural marshmallows?"[4] They are hard to resist (as we will see later).

PAPER OR PIXELS?

But this chapter is about book reading, and we need to crack the spine and dig into a big debate. What's better for reading: paper or pixels?

To answer this question, two psychologists compared digital and print comprehension using a short article of about a thousand words. Participants were split into two groups. One group read the article on a screen and the other group read it from paper. They were required to invest an identical amount of time to read the piece. Then the experiment was conducted a second time with new groups, but with one key difference: the time frame for reading

---

4. Matthew Crawford, *The World beyond Your Head: On Becoming an Individual in an Age of Distraction* (New York: Farrar, Straus and Giroux, 2015), 16–17.

was removed, leaving readers to set their own pace. Finally, all the groups were tested for retention. The participants in the first experiment, asked to invest identical reading time whether they read on screen or paper, scored nearly the same in the test. But in the second experiment, print readers noticeably outscored their digital counterparts. Why? The reason was simple: digital readers naturally read too quickly.

The takeaway in the study was simple and yet profound: poor digital reading was not a result of the medium, "but rather of a failure of self-knowledge and self-control: we don't realize that digital comprehension may take just as much time as reading a book."[5] With digital text on our phones, we are conditioned to skim quickly. With a printed book in hand, we naturally read more slowly, at a pace realistic for retention. Simply put, "If you want to internalize a piece of knowledge, you've got to linger over it."[6] But we have been trained to *not* linger over digital texts.

The digital age hurries us and shatters our concentration into a million little pieces, says ethicist Oliver O'Donovan, and now the greatest challenge to literacy is a short attention span, "caught now by one little explosion of surprise, now by another. Knowledge is never actually given to us in that form. It has to be searched for and pursued, as the marvelous poems on Wisdom at the beginning of Proverbs tell us."[7] And it is always wise to contrast our social-media habits with the disciplined wisdom-seeking habits celebrated in the first three chapters of Proverbs. Our lack of self-control with digital marshmallows malnourishes our sustained linear concentration.

Deep reading is harder than ever. Today, given the amount of written words that come into our lives in a given day, we have simply grown careless. "Once I was a scuba diver in the sea of words," laments writer Nicholas Carr. "Now I zip along the surface like a guy

5. Maria Konnikova, "Being a Better Online Reader," *The New Yorker* (July 16, 2014), summarizing Rakefet Ackerman and Morris Goldsmith, "Metacognitive Regulation of Text Learning: On Screen versus on Paper," *Journal of Experimental Psychology: Applied* (March 17, 2011), 18–32.

6. Clive Thompson, *Smarter Than You Think: How Technology Is Changing Our Minds for the Better* (New York: Penguin, 2013), 135.

7. Oliver O'Donovan, interview with the author via email (Feb. 10, 2016).

on a Jet Ski."[8] We can hardly get submerged into the serious work of reading a book before we feel the desire to digitally surface and Jet Ski (skim) over easier waters.

But whatever the cause, the literacy problem we face today is not *illiteracy* but *aliteracy*, a digital skimming that is simply an attempt to keep up with a deluge of information coming through our phones rather than slowing down and soaking up what is most important. Those who are *aliterate* have difficulty separating what is eternally valuable from what is transient. They skim, but not in order to identify and isolate what needs to be studied more carefully and meditatively. Because the *aliterate* cannot navigate this distinction, they struggle to draw relevance from written texts, especially ancient texts.

Will we eventually train ourselves to read digital texts more slowly and carefully? That answer is unknown. We do know that when it comes to digital texts today, we tend to skim in unnaturally rushed and anxious speeds. We aren't very good at lingering over digital text.

COVENANTAL CONCENTRATION

Digital reading is unnecessarily hurried, and this habit bleeds into how we read our Bibles. Hip-hop artist and pastor Trip Lee admitted to me something I think we've all experienced: "The more time I spend reading ten-second tweets and skimming random articles online, the more it affects my attention span, weakening the muscles I need to read Scripture for long distances."[9] But before we delete our Bible apps, we should consider that studies also tell us that Christian readers are more faithful to follow digital Bible reading plans on smartphones (with daily prompts) than print plans and offline reading.[10]

So whatever the medium (paper or pixels), and whatever the

---

8. Nicholas Carr, *The Shallows: What the Internet Is Doing to Our Brains* (New York: W. W. Norton, 2011), 7.

9. Trip Lee, interview with the author via Skype (March 25, 2015).

10. John Dyer, "Print Bibles Vs. Digital Bibles: Comparing Engagement, Comprehension, and Behavior," unpublished draft (March 2016). His study also confirms the findings of Ackerman and Goldsmith.

weakness of the medium's users (forgetfulness or hastiness), we must become mindful and slow our pace. The Bible is a covenant document from God to us. It spells out the relationship we enjoy with him, teaches us the blood-bought promises he has made, and instructs us how to live in this world to display our covenant faithfulness to him.

"The commerce and communion between God and his people is an inherently textual phenomenon. The eternally eloquent God has stooped to speak a word of saving consolation to us," writes theologian Scott Swain. "Because Scripture is the supreme locus of God's self-communication in the world, Christians are 'people of the book.' The Lord gathers, nourishes, defends, and guides his people through this book; and his people assemble around, feed upon, find shelter in, and follow after the words of this book."[11]

God has given us the power of concentration in order for us to see and avoid what is false, fake, and transient—so that we may gaze directly at what is true, stable, and eternal. It is part of our creatureliness that we are easily distracted; it is part of our sinfulness that we are easily lured by what is vain and trivial.

Our joy in God is at stake. In our vanity, we feed on digital junk food, and our palates are reprogrammed and our affections atrophy. "To be sure, the compulsive shopper, or gamer, or Facebooker, may be trying to fill the God-sized hole in their life, or to drown out his summons with a white noise of frenetic triviality. But as with all vices and virtues, there is something of a feedback loop at work here," explains Brad Littlejohn, an independent scholar. "The more we take refuge in distraction, the more habituated we become to mere stimulation and the more desensitized to delight. We lose our capacity to stop and ponder something deeply, to admire something beautiful for its own sake, to lose ourselves in the passion for a game, a story, or a person."[12]

By seeking trivial pleasure in our phones, we train ourselves to

---

11. Scott R. Swain, *Trinity, Revelation, and Reading: A Theological Introduction to the Bible and Its Interpretation* (London; New York: T&T Clark, 2011), 95.

12. Brad Littlejohn, "The Seven Deadly Sins in a Digital Age: 4. Sloth," Reformation 21, reformation21.org (November 2014).

want more of those trivial pleasures. Most seriously of all, "either we, out of fear and guilt, lose our delight in God, the source of all good, and thus begin to lose our delight in all the goods he has given us, till we care less and less for anyone or anything, and lose ourselves in momentary diversions, which then become the only 'pleasures' we know—or we begin to thoughtlessly habituate ourselves to the ecosystems of distraction that surround us, until we begin to forget what it might feel like to truly attend to a poem or a person," Littlejohn says. "Our capacity for deep enjoyment thus destroyed, we quickly lose the capacity to enjoy the One who demands the most sustained attention of all."[13]

BIBLE CONCENTRATION

These heart consequences land heavily, but this is where cosmic purpose meets personal discipline. We are called to suspend our chronic scrolling in order to linger over eternal truth, because the Bible is the most important book in the history of the world.

It can be said that literacy has fallen to such a degree that, for many Christians today (perhaps *most* Christians today), the Bible stands as the oldest, longest, and most complicated book we will ever try to read on our own. Simultaneously, every lure and temptation of the digital age is convincing us to give up difficult, sustained work for the immediate and impulsive content we can skim.

Bible reading is incredibly demanding work, yet I find much comfort and hope in knowing the Bible calls me to lifelong engagement. The Bible is not a book to "get through," to read cover to cover and then put on a shelf; neither is it a book to browse or skim. The Bible is our open door to hear God's voice both alone and together in community. It is intended to be bottomless in its profundity and endless in its relevance. It is less of a book and more of a world of revelation in which we live and move and have our being. This book gives us life, and it moves and pushes God's redemptive plan forward.

---

13. Ibid.

In fact, "the whole purpose of God for the universe stands or falls on the book. If the book fails, everything fails."[14]

So to skim the Bible is to misread it, points out New Testament theologian Daniel Doriani in three points. First, the aim of the Bible is discipleship, to continually form and re-form our thinking, our habits, and our behaviors. This dynamic process never ends, and thus our reading never ends—and there's no benefit to skimming to the end. Second, the Bible's Author warns us over and over again that the book will be rejected, distorted, or misunderstood in various ways. Stern internal warnings caution us to slow down and read with care, prayer, precision, and urgency.[15] Third, "the Bible's Author and authors have chosen to reach their goals not by straightforward lecture, proceeding proposition by proposition, but through songs and poems, dark sayings, and half-interpreted stories." In other words, "We readers don't take dictation; we swim in metaphor."[16] And to appreciate those metaphors, we must deep-dive into the divine text for a lifetime.

(SELF-)CONTROL FREAKS

It is in the slow reading of the ancient Bible that we feed on the full benefit of writing, says O'Donovan. "Ephemeral text [of the digital world] does not represent the distinctive strength of textual communication, which is its power to cover distance, to open up historical and local views not accessible to immediate exchange."[17] Social media are far too new, too contemporary, too close, too much like me to tap into the greatest benefit of literacy. To be changed and challenged, we need the clean sea breeze of old books, said C. S. Lewis.[18] We need the life-living gust of the Spirit in the ancient book. And when it comes

---

14. John Piper, in a personal conversation (March 18, 2016). Shared with permission.

15. Examples include 2 Pet. 3:15–16 and the "have you not read" statements from Jesus (Matt. 12:3–7; 19:4; 19:4–5; 22:31).

16. Daniel M. Doriani, "Take, Read," in *The Enduring Authority of the Christian Scriptures*, ed. D. A. Carson (Grand Rapids, MI: Eerdmans, 2016), 1123–24.

17. Oliver O'Donovan, *Ethics as Theology*, vol. 2, *Finding and Seeking* (Grand Rapids, MI: Eerdmans, 2014), 133.

18. C. S. Lewis, *God in the Dock* (New York: HarperOne, 1994), 220.

to serious literacy, the faithful church is counterculturally positioned for success, because solid expositional preaching is essentially a model of healthy, slow reading.[19]

In the smartphone age, we are bombarded daily by the immediate: Facebook updates, blog posts, and breaking news stories. Yet the most important book for our soul is ancient. God's Word demands our highest levels of literary concentration because it requires relational reading: not the superficial chitchat of a cocktail party, but the covenantal concentration of marriage vows. God's Word is an invitation to orient our affections and desires.[20] Our challenge is to use social media in the service of serious reading.

So what sort of freaks of self-control must we become to resist the well-engineered marshmallows of distraction? Freaks who believe in 2 Corinthians 4:18, who "look not to the things that are seen but to the things that are unseen. For the things that are seen are transient, but the things that are unseen are eternal." And this challenge leads us to our next stop.

---

19. C. Christopher Smith, *Reading for the Common Good: How Books Help Our Churches and Neighborhoods Flourish* (Downers Grove, IL: InterVarsity Press, 2016), 27–28.

20. Psalm 119 is a long and prolific chapter about obedience, and it's loaded with the language of *heart fullness, delight, joy, awe, praise,* and *singing.* The key to obedience is not simply reading God's law, but having a heart filled with delight in the Lawgiver and his words to us. Our defense against sin is a heart full of God-centered affection.

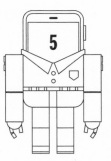

WE FEED ON THE PRODUCED

Fish live in water. Celebrities live in replicating images.

For celebrities to survive another day, they must find ways of replicating images of themselves over and over. Celebrities must stay in the news—that is their job—and the corporations that bank on the celebrities need to keep pushing those icons forward, too. This means celebrity culture survives on cameras—lots and lots of cameras: still cameras, video cameras, studio cameras, paparazzi cameras, and fan cameras.

Not only do our smartphones have sharp cameras that capture quality images and video, those cameras are always with us, and we have developed fidgety "shutter" trigger fingers, ready to capture anything in the moment. Taken together, we not only consume celebrity culture, we now feed the culture, too.

ONE ICONOGRAPH

When it was announced that megastar Johnny Depp and his fellow cast members would show up for a movie premiere in Boston, fans piled in tight against the makeshift metal fencing flanking the red carpet along a sidewalk leading up to a closed-off theater. As Depp and various other actors and actresses appeared on the red carpet, hundreds of cameras ignited. In a moment of genius and paradox, one seasoned Boston photographer turned his attention from the stars to the tightly packed crowd of onlookers and snapped a photograph that is an icon of our age.[1]

In the frame, I see forty-four onlookers tightly pressed together and at least thirty visible smartphones raised up in the air, cameras on. One middle-aged man in the front row fidgets with an app, no doubt trying to get his camera to work. Almost everyone else is ready for the moment, holding phones straight up as high as their arms will go to get the clearest possible picture or video of the procession. This means almost everyone in the picture is looking away from Depp, gazing upward into phones in a comical posture future generations will most certainly enjoy mocking.

But foregrounded among the throng of raised smartphones stands one elderly woman, her arms leisurely folded across her chest and resting on the top railing of the barricade. She looks directly at the actors with a carefree repose and a small grin. She's not trying to capture or share anything, not trying to grab a picture or moving frames to share online later. She is simply enjoying the moment.

To her left stands a younger woman who holds her phone out to record the scene, but whose eyes are firmly fixed on the event before her, not on her screen. Unlike the others, she has the wherewithal to hold her phone, but also to enjoy the moment with her own eyes.

ENJOYING THE MEDIATED

The crowd represents a spectrum of attitudes in the moment (just watch, watch and capture, just capture), and we need to know our own bent here. But before returning to smartphone camera compulsions, we need big-picture definitions, and for that I'll go pano for a moment.

First, and most foundational, everything in this world that we

---

1. Assistant chief photographer John Blanding of *The Boston Globe* took the photo on September 16, 2015. To see the image, see Emily Anderson, "This Boston Globe photo is perfect," BDCwire, bdcwire.com (Sept. 28, 2015). Depp was the lead actor in *Black Mass* (2015; R), a movie about Whitey Bulger (b. 1929), the convicted murderer and infamous mob boss in South Boston.

hear, see, smell, touch, and taste exists because God spoke it into existence. He spoke—and all of the visible and observable creatures, features, and forces came into being.[2] He spoke—and light, animals, oceans, mountains, sunrises, full moons, forests, and a spectrum of paint colors came into being. What he spoke into existence, he continues to speak through, calling human worshipers to delight in him as we enjoy what he made.

All of creation is a footpath back to God. So if you look at the burning sun and ask, "What is the sun?" the full answer is not a perpetually exploding atomic bomb of volatile gases. As Christians, we press past the physics and ask: "Yes, but *why* does the sun exist in the first place? Who put the sun into space? And what does that tell us about the *person* who conceived and made the sun come into reality?"

Nothing in the universe is arbitrary or coincidental, because nothing exists by random chance. All things come from someone, are caused by someone, and thus contain meaning beyond the nature of themselves. "For from him and through him and to him are all things. To him be glory forever. Amen" (Rom. 11:36). For those with eyes to see its true meaning, the burning sun in the sky is a display of God's glory, and it serves as a placeholder for a greater glory to be revealed later.[3] So everything real finds its origin in God—meaning that all of creation is *mediated*.

But add another level, a middleman—and now we are talking about *intermediated* experiences. Everything we read, hear, see, or watch on our phones falls under this category. On the screens of our smartphones, we find only copies of what exists in the world. We read messages only as they are intermediated to us by others, by the gatekeepers of the creative world—from musicians, artists, movie producers, and even our friends and family members.

This may not make sense yet, so let me stack *mediated* and *intermediated* together in three distinct categories:

---

2. Gen. 1:1–31; Heb. 11:3.
3. Rev. 22:5.

Natural Revelation from God (Mediated, Part 1)

God, who is invisible, spoke creation into being[4] and now mediates his presence to us through his material creation and the natural laws he sets in motion, which, originally perfect, are now tainted and fallen[5] but never muted.[6]

Special Revelation from God (Mediated, Part 2)

God, who is invisible, speaks himself to the universe in his words, his works, and, ultimately, in the incarnate Word, Jesus Christ. Christ is the definitive Word of God, and he is made known to us now in the pages of Scripture.[7]

Productions of Man (Intermediated)

Image bearers of God (us) take the materials of the world, and the words, natural laws, and values of human flourishing established by the Creator, and re-mediate all of it through our cultural products—art, music, literature, and texts—adding a layer of interpretation in our creations, intentionally or inadvertently, for good or evil.

God's words and works always precede man's words and works. God spoke creation into existence and spoke his definitive word to us in the person of his Son, Jesus Christ. God ordered the world, wisdom, and redemption, and he set the stage for human art.

Here's another way to say it. God created us in order to shower his gifts over our lives, beginning with the natural wonders of breath, sunshine, food, water, rain, beaches, and mountains. As we receive these gifts (and many others), we stop at key moments to respond to him with joyful thanks.[8] He must break the power of sin for this gratitude to work properly in our lives, but when it does, we are given the gift of God-centered thankfulness to embrace his natural order, to receive all of his cosmic wonders, to enjoy the "thickness" of his

---

4. Gen. 1:1–31; Heb. 11:3.
5. Gen. 3:14–19; Rom. 8:18–25.
6. Rom. 1:18–23.
7. John 1:1–18; 2 Cor. 4:4–6; Col. 1:15–20; 2:1–8; Heb. 1:1–3.
8. Rom. 1:18–23; 1 Tim. 4:1–5.

material gifts,[9] and to delight in our friends and spouses—receiving from God our entire existence: our lives, our lots, our souls, our bodies, our biological genders, and his astonishing but unblushing design for human sexuality and procreation.[10]

The Spirit makes us spiritually alive in order for us to see God's glory revealed in Scripture, and the Spirit opens our eyes to see the Creator who stands behind all our natural gifts.[11] The Bible reveals to us a God who is eager to bless us physically and spiritually—and the supreme proof of his brimming generosity is the shed blood of his precious Son.[12] As we behold Christ's glory by faith in Scripture, our hearts swell in thanksgiving to greater heights.[13] All things that we have been given—and all that we are or ever hope to become—are gifts in Christ, who is both our Creator and Redeemer.[14] In Christ, we see through the showers of gifts to behold the glory of the Giver as we wait for an eternity in the matchless delights of God's presence.[15] He is the supreme gift of all—the gift toward which all the other gifts have been pointing and leading us all along!

---

9. Douglas Wilson: "Creation is a gift meant to bring glory to the Creator. All Christians agree here. But Christians throughout the ages have put their suspicions in different places. Take C. S. Lewis and Augustine. I love them both, but I would rather have a beer with Lewis. Lewis would order us a really good beer, just because it was a really good beer, with his understanding of God suffusing the whole. For him, while the thickness of creation *can* become an idol, a rival to God, it is intended for us as a sermon by God about God. And you can't honor the preacher by ignoring the sermon. But Augustine would perhaps think that a thin beer would help us think of Jesus more, not distracting us quite so much, and that when we had really advanced in grace, we might be able to get the same effect with water. I say this in the full recognition that I am not worthy to have been Augustine's boot boy. So then a right approach to a thick creation honors the Creator more fully; we honor his work as he gave it, instead of trying to dilute it in a misguided zeal for his glory." Email to the author (July 1, 2016). Shared with permission. Intended or not, Wilson's illustration of alcohol density coincides with the display of divine glory echoed in the first miracle of Jesus (John 2:1–11). He did not flex his sovereign power by turning party wine into water, but by turning ceremonial washing water into dark, undiluted party wine—the "good wine" that caught attention. Not only did the water-to-wine thickening of creation *not* cloud Christ's glory, it manifested it. See also Douglas Wilson, "Creation Is Thick, I Tell You," *Blog & Mablog*, dougwils.com (May 16, 2010); and Joe Rigney, *The Things of Earth: Treasuring God by Enjoying His Gifts* (Wheaton, IL: Crossway, 2014), 74, 95–115.

10. Psalm 16; Prov. 5:18–19; Eccles. 9:9.

11. Herman Bavinck: "If we had not heard God speaking to us in the works of grace and by that means also discerned his voice speaking to us in the works of nature, we would all be like pagans, for whom nature speaks in a cacophony of confusing tongues." *Reformed Dogmatics*, vol. 2, *God and Creation* (Grand Rapids, MI: Baker Academic, 2008), 75.

12. Rom. 8:32.

13. 1 Corinthians 2.

14. John 1:3, 10; 1 Cor. 8:6; Col. 1:16; Heb. 1:2.

15. 1 John 3:2; and Psalm 16 again.

So should we pass our days in silent longing as we wait for this visible glory? No, we cannot. By faith, we must boast in Christ now, our Savior and Maker and the sustainer of all things.[16] Our souls have been raised to new life in order to brag of Christ, and as we speak, our joy expands and overflows, and we become creators and artists. Art is spontaneous. Art is doxology. Art is the reflection of God's beauty into the world. This is why we exist!

In speaking of the purpose of our lives, we've gotten ahead of ourselves (more in the next chapter). Let's speak here of the purpose of our media.

Every artist works only with the raw materials of God's generosity, and this leads to two conclusions.

Negatively, to express godless art means that no higher purpose exists than the fame of the artist. Godless art does not merely shrug off God or innocently forget him—godless art actively prybars God from his created reality and suppresses the reflection of his glory with a thick layer of black paint.[17]

Positively, to express Christ-honoring art means that everything we create, share, and spread on our phones—paintings, music, photography, poems, and books—can amplify God's natural and special revelation. So we aim to produce art that reflects God's glory in undiminished splendor.

In either case, in everything we make, we add a layer of interpretation. So I must always ask myself, does my digital art dim glory or reflect glory?

JEEP CALLS TO JEEP

Here's one simple example. Imagine you open Facebook and see a beautiful drone video of the Grand Canyon. You take a moment to watch the cinematic clip that captures something of the depth and scope of the majestic scene, all complemented by poetic narration and haunting music. It's breathtaking.

---

16. Rom. 11:36; 1 Cor. 8:4–6; Gal. 6:14.
17. Rom. 1:18–32.

This video can be presented to us with starkly differing aims. First, it can be used to stir worship in us as we see the majesty of God's natural glory re-mediated through human production. Or this same footage can be used to stir our love for a new product, such as an off-road Jeep. One interpretation amplifies the glory of God, while the other amplifies the craftiness of a corporate marketing firm.

In either case, watching the video is no match for standing on the lip of the Grand Canyon and beholding creation directly as an encounter with God's vastness. The video cannot fully spark this awe.

My point is simple. We must be aware that all the content on the "small screen" of our phones is intermediated. This is not good or bad, just a reality that calls for discernment and discretion. On our phones, we have high-definition portals into the vast beauties and glories of creation, but every message we receive has been cut, edited, and produced for a purpose. This distinction also keeps our smartphone screens in proper context when it comes to God's massive glories—seen and unseen—that surround our lives.

POINT, SHOOT, FORGET

Now, I've managed to open up a huge discussion about art, but we are only talking about smartphones, so to keep this chapter short, I will now compress some key points and mostly point out a few suggestions and implications, first by returning to the topic of our smartphone cameras.

The high-resolution cameras built into our phones are simply one of the most incredible blessings of the digital age—convenient, portable, and potent. But they also raise three questions.

First, we need to think about the social capacity of our phones and how that capacity shapes our impulses. What is true of our cameras is true of every smartphone behavior—the power to immediately share anything we see or do conditions what we capture in the first place. In Donna Freitas's extensive study of the social-media habits of college students, one sharp female student told her: "People used to do things and *then* post them, and the approval you gained from whatever you were putting out there was a byproduct of the actual activity. Now the *anticipated* approval is what's driving the behavior or the activity, so there's just sort of been this reversal."[18] Phones with social connections transform us—and our friends and children—into actors. That's huge.

Second, we need to rethink our memories. What if the point-and-shoot cameras in our phones make us less capable of retaining discrete memories? One psychologist calls this camera-induced amnesia the "photo-taking impairment effect,"[19] and it works like this: by outsourcing the memory of a moment to our camera, we flatten out the event into a 2-D snapshot and proceed to ignore its many other contours—such as context, meaning, smells, touch, and taste.

If the cameras in our pockets mute our moments into 2-D memories, perhaps the richest memories in life are better "captured" by our full sensory awareness in the moment—then later written down in a journal. This simple practice has proven to be a rich means of preserving memories for people throughout the centuries. Photography is a blessing, but if we impulsively turn to our camera apps too quickly, our minds can fail to capture the true moments and the rich details of an experience in exchange for visually flattened memories. Point-and-shoot cameras may in fact be costing us our most vivid recollections. But until we are convinced of this, we will continue to impulsively reach for our phones in the event of the extraordinary (or less).

Third, and most insidious of all, I wonder if this unchecked impulse exposes something deeper and darker in us, a certain unbelief that drives us, something more similar to the lie that maybe a given moment is our last opportunity to get close to greatness. In essence, this was the scam that targeted Adam and Eve, and it has been the heart of every human dupe ever since.[20]

Sin lies about the future. If I don't grab this chance at glory now,

---

18. Donna Freitas, *The Happiness Effect: How Social Media Is Driving a Generation to Appear Perfect at Any Cost* (New York: Oxford University Press, 2017), 4.

19. Jeff Jacoby, "Free Your Eyes from the Shackles of the Shutter," *The Boston Globe* (Oct. 4, 2015).

20. Gen. 3:4–5.

sin tells me, it will be lost forever. So we point our phones at celebrities, which only points out our forgetfulness. We forget eternity. We so easily lose the faith to imagine that one day we will inherit the world and be more renowned and wealthier than Johnny Depp could ever imagine in this life.[21] We want our share of glory now, instead of waiting for our "glory that is to be revealed."[22] What if our rhythms of Snapchat selfies and our star-studded Instagram feeds are exposing the dimness of our future hope?[23]

BREAKING FREE

How, then, can we walk (and click and share) with wisdom?

First, we must humbly admit that we are targets of digital mega-corporations that can make us into restless consumers with strategic intermediated content. We cannot be naive here. Our attention spans have been monetized, and getting us hooked on our phones is a commercial commodity measured in billions of dollars, not in kiosk change. The hook often comes in visual allurements. Again, this medium is not inherently wrong. Digital art and messaging can be done for God's glory, and done well. But we must see that we are being conditioned to turn to our phones when we want to be amazed and wowed, and in turn, we are being milked for corporate profit. Likewise, social-media platforms are huge businesses with public stock prices, and they can grow in value only if they condition us to become actors in front of our phones.[24]

---

21. Ps. 37:11; Matt. 5:5; 25:21; 1 Cor. 3:21–23; 2 Tim. 2:12; James 2:5; Rev. 2:26; 5:10.
22. Rom. 8:18; 1 Pet. 5:1.
23. Phil. 3:19.
24. Entrepreneur Seth Godin: "Social media wasn't invented to make you better, it was invented for you to make the company money. By it you become an employee of the company. You are the product they sell. And they put you in a little hamster wheel and throw treats in now and then. . . . The big companies of social media went from being profoundly important and useful public goods that created enormous value, to becoming public companies under pressure to make the stock price go up." Tim Ferriss, *The Tim Ferriss Show* podcast, "How Seth Godin Manages His Life—Rules, Principles, and Obsessions," The 4-Hour Workweek, fourhour-workweek.com (Feb. 10, 2016). To develop the analogy further, the hamster wheel is also a cogwheel, with its teeth locked into the cogs of other hamster wheels. As long as one hamster runs, all the other wheels begin to turn, obligating all the other hamsters to run too. The power of social media (its interconnectedness) generates the torque of one interlocked machine that, once started, may move at varying speeds, but becomes inexorable. The machine will not stop. All the hamsters must run.

Second, we must learn to enjoy our present lives *in faith*—that is, to enjoy each moment of life without feeling compelled to "capture" it. A growing trend among touring musicians is to ask fans not to film concerts on their phones. Keep the phone in your pocket and enjoy the moment, they say. This direction parallels something of the Christian enjoyment of God's good gifts. Get off your phone, go camping, gaze at the stars, hike in nature—whatever brings creation closer and richer than pixels.

Third, we must celebrate. We cannot suppress our souls' appetite for what is awe-inspiring. The goal is not to mute all smartphone media but to feed ourselves on the right media. We were created to behold, see, taste, and delight in the richness of God's glory—and that glory often comes refracted to us through skilled artists. Our insatiable appetite for viral videos, memes, and tweets is the product of an appetite for glory that God gave us. And he created a delicious world of media marvels so that we may delight in, embrace, and cherish anything that is true, honorable, just, pure, lovely, commendable, excellent, or worthy of praise.[25] This will keep us very busy marveling at Scripture, at nature, and at God's grace in the people he created.

FEEDING AUTHENTICITY

Filled with *mediated* reality from God, we become eager in our celebration and shrewd in our discernment of *intermediated* art. For our online networks, we become filters—salt and light—as an act of love in what we publish, share, and like. We refuse to be brainless carriers of the most recent viral meme. Instead, we live as Christians offering "dialogical resistance"—which means that we filter the messages of the world through our individual discernment and then share online through a robust theology of reality, possibility, and meaning in God.[26]

To do this, we must escape the trap of the intermediated world of the produced and step away to live our own lives. On the nine-

---

25. Phil. 4:8.
26. Oliver O'Donovan, *Ethics as Theology*, vol. 2, *Finding and Seeking* (Grand Rapids, MI: Eerdmans, 2014), 83, 87.

month anniversary of her social-media sobriety—completely off Instagram, Pinterest, Facebook, and Twitter—my wife turned to me and said, "Compulsive social-media habits are a bad trade: your present moment in exchange for an endless series of someone else's past moments." She's right about the cost. Our social-media lives can stop our own living.

Or, as Andy Crouch says, our smartphone addiction leads to creational blindness. It is only in the absence of constant digital flattery that we can feel small and less significant, more human, liberated to encounter the world we are called to love.[27] We inevitably grow blind to creation's wonders when our attention is fixed on our attempt to craft the next scene in our "incessant autobiography."[28] Instead, says Crouch, "All true, lasting creativity comes from deep, risky engagement with the fullness of creation." So "get out in the glorious, terrifying creation and let it move you and break your heart. Then you'll have something to offer in the dim mirror that is 'social media'—and in the full, real world that demands the engagement of all of our heart, mind, soul and strength."[29] Yes, step away from screens, and let the glories of creation break your heart and let the handiwork of God's creative genius wash you as you ski mountains, hike trails, and scuba dive into oceans. But don't stop there. Climb the summits of Scripture, too. Let God's Word pierce your intentions and cut down into your truest motives, and let yourself be convicted, broken, and remade—which is the feeling of standing in the breathtaking presence of God.[30]

Then take all of God's created and revealed gifts to you and make all of them into a life that shows the world how glorious and satisfying God really is. This is the secret to "creating" great digital art of all forms and types.

---

27. Andy Crouch, "Small Screens, Big World," *Andy Crouch*, andy-crouch.com (April 8, 2015).

28. C. S. Lewis's summary of Satan's driving motive in John Milton's epic poem *Paradise Lost*. Having traversed heaven, hell, and the whole cosmos, Satan finally becomes focused only on himself—an infinite boredom inescapable. Adam, born into a small park, is so quickly filled with awe and wonder at the creation that he seems to almost forget himself in the grandeur of it all. See C. S. Lewis, *A Preface to Paradise Lost* (London: Oxford University Press, 1961), 101–3.

29. Joshua Rogers, "Five Questions with Author Andy Crouch," *Boundless*, boundless.org (June 15, 2015).

30. Heb. 4:12–13.

CALLING ALL ARTISTS (AN EXCURSE)

The chapter has now ended, and it was mainly talking to content consumers, but I must speak more specifically to serious digital artists (of all skill levels), and this seems like a good place. So begins a sidebar for artists, sharers, and creators.

Christians who create and share digital media have more open doors and opportunities for expression than at any other time in church history. The posture of the church is not tilted backward and away from digital media, but forward and open to new uses of technology—to the production of articles, poetry, spoken word, music, movies, vlogs, podcasts, novels, photography, and paintings, all for the purposes of reflecting God's glory, engaging the world with a biblical worldview, and even proclaiming the hope of the gospel.

Jesus's metaphor for gospel labor is a farmer who tosses seeds all over the ground, hoping some of those kernels will take root, grow, and flourish into a crop.[31] In the same way, Christian leaders and artists are called to broadcast truth all over the place, prayerfully hoping some of it will take root in hearts. I am not advocating cheesy religious memes, but deep, thoughtful, original reflections that emerge from the place where creation and biblical truth meet your life and worship. Christian artists express this personal intersection with unique, expressive gifts. And we all have the tools. Everyone with a phone is not merely a content consumer, but now a producer *and* a consumer—a *prosumer*, as they say. All of us are apologists, teachers, advocates, and prophets, speaking into the lives of others. These new modes of cultural expression, sharpened in the hands of discerning artists, become strategic gospel weapons.

This astounding opportunity presses one big question: What is the ultimate purpose of my art? Technology is prag-

---

31. Mark 4:1–20.

matic; it presses us to ask *how*, not *why*.[32] The mechanisms and techniques of technology naturally trump the questions about ultimate aims. So, in the digital age, we must ask this question of purpose over and over.

Christians who ask this question ("Why do I create art?") find that "self-expression" is an insufficient answer. We are commanded to constant self-evaluation of every behavior and practice, all gauged by ends, aims, and goals. The apostle Paul gives us a gold standard of Christian ethics in one ancient back-and-forth debate.

"All things are lawful for me," came the reigning motto from ancient Corinth.

"But not all things are helpful," responded Paul.

"All things are lawful for me," came the motto again, perhaps louder.

"But I will not be dominated by anything," retorted Paul.

"All things are lawful," came the motto a third time, now even more abruptly.

"But not all things are helpful," reiterated Paul.[33]

Freedom in Christ is not freedom to do whatever you want; it is for sure-footed self-reflection and for avoiding the cultural bondage of sin. My freedom in Christ gives me eyes to see that not all things are helpful for me, helpful for others, or acceptable for my witness in the world.

In principle, Paul continually presses Christian creators to ask three questions:

- Ends: Do my art and social media point others toward God?
- Influence: Do my art and social media serve and build up my audience?

---

32. See Langdon Winner, *Autonomous Technology: Technics-out-of-Control as a Theme in Political Thought* (Cambridge, MA: MIT Press, 1977).

33. 1 Cor. 6:12–13; 10:23.

- Servitude: Do my art and social media imprison me into an unhealthy bondage to my medium?

The Weight of My Words on Others

These principles hold true in everything we create, but especially with the words we craft. Even our digital words should point others Godward. "Idle words" cannot do this, so Jesus tells us that "every careless word" should be put off.[34] We must do away with words that destroy.

To use one striking biblical metaphor, our tongues are like fire—capable of blessing as well as destroying.[35] With our tongues, we bless God and curse God's image bearers. An untamed tongue is like a wheel on fire, rolling along its course and spreading flames as it goes. We are constantly pushing and shoving the trajectories of one another by our tongues (through our thumbs).

Perhaps C. S. Lewis's most prophetic word to the digital age takes up this theme. "It is a serious thing to live in a society of possible gods and goddesses," he wrote, not elevating man as gods, as the Serpent did with his lie, but as gods and goddesses in their glorified state in the new creation. Maybe we can imagine being glorified ourselves, but we often lack this imagination for our neighbors. In fact, "the dullest and most uninteresting person you can talk to may one day be a creature which, if you saw it now, you would be strongly tempted to worship, or else a horror and a corruption such as you now meet, if at all, only in a nightmare."[36]

To be sure, every human will stand before God to give an account for his or her life and bear the eternal weight of his or her faith or unbelief. But it also remains true that every day we are leading each other in one of two directions: (1) toward Christ

---

34. Matt. 12:36.
35. James 3:1–12.
36. C. S. Lewis, *The Weight of Glory: And Other Addresses* (New York: HarperOne, 2001), 45.

and an eternal beauty that will one day take our breath away or (2) toward rejection of Christ and an eternally distorted ugliness and soul decay, reminiscent of the evil only barely hinted at in modern horror films. "It is in the light of these overwhelming possibilities, it is with the awe and the circumspection proper to them, that we should conduct all our dealings with one another, all friendships, all loves, all play, all politics"—and all our social media. "There are no *ordinary* people. You have never talked to a mere mortal. Nations, cultures, arts, civilizations—these are mortal, and their life is to ours as the life of a gnat. But it is immortals whom we joke with, work with, marry, snub, and exploit—immortal horrors or everlasting splendors."[37]

Of course, this warning relates directly to cyberbullying, but the principle extends to all of our texts and tweets. Behind the words in our mouths we find desires in our hearts, and those desires are always sparking new desires in the hearts of others.[38]

In summary: "the people you text and tweet," said David Platt, "are going to spend the next quadrillion years either in heaven or hell."[39] And his timeline is understated. Sticks and stones may break bones, but my texts and tweets are pushing eternal souls in one of two directions. Let this sobering truth guide your art.

The Weight of My Words on Me

Here's where it also turns personal. When it comes to our language, Jesus warns us that we speak from what's already stored up in our hearts: "What comes out of the mouth proceeds from the heart, and this defiles a person" (Matt. 15:18). The heart origin of our words alone is a good check on our social-media

---

37. Ibid., 46, emphasis original.
38. On the potency of mimetic desire, see René Girard, *Deceit, Desire, and the Novel: Self and Other in Literary Structure* (Baltimore: Johns Hopkins Press, 1965), and *Theater of Envy: William Shakespeare* (New York: Oxford University Press, 1991).
39. David Platt, sermon, "The Urgency of Eternity," Radical, radical.net (March 10, 2013).

use, because the typed words of our thumbs manifest the core loves and desires of our hearts. But what haunts me even more is the second half of the verse, where Jesus makes it clear that our words don't merely expose us, they define us; and not only do they define us, they can destroy us.

Jesus echoes a paradigm found all over the Bible: "Whoever guards his mouth preserves his life; he who opens wide his lips comes to ruin" (Prov. 13:3); "The words of a wise man's mouth win him favor, but the lips of a fool consume him" (Eccles. 10:12). Note the incarnation of our words. Over and over we are warned that every time we speak, we birth words into the world. We speak a legacy. Our words linger around us, they grow in power, and they either improve us or—like uncontrolled fire—turn against us. If we are self-controlled, the words we use to build others will also build us. But if we lack self-control, the unfiltered digital words we speak through our phones will be like an army spat from our mouths that will make war on us and damage our lives in every way—relationally, socially, financially, physically, and spiritually.[40]

"Death and life are in the power of the tongue" (Prov. 18:21). With our digital words, we can destroy others and we can destroy ourselves. With our digital words, we can build up others and we can bless ourselves. The profound biblical link is key. Our words destroy us if they are meant to destroy others, but our words build us up if they are meant to bless others. This means that for most of us, with our modest social-media platforms, the greatest influence of our smartphone words will be found in the power and influence they wield over us.

Wielding Digital Words

Lightning-fast distribution of digital words, music, and images is a tremendous tool, but it also requires skill, a skill we

---

40. Pss. 64:8; 140:9; Prov. 10:14; 12:13; 13:3; 14:3; 18:6–7, 20–21; Eccles. 10:12–14.

can learn by returning to a threefold ethical paradigm for all of Christian living: (1) kill the sinful habits of life that misuse God's good gifts while (2) magnifying the Giver for the gifts themselves by (3) employing the gifts with missional purpose. In this case, replace "gifts" with "gifts of digital media." Kill the sinful habits of life that misuse God's good gifts of digital media while praising the Giver for the gifts of digital media by employing digital media with missional purpose.

If everything we post on Facebook, Twitter, and Snapchat, and everything we write in our text messages and emails, is produced, we ourselves are producers. So *what* am I producing, and, more important, *why* am I producing it? Before you text, tweet, or publish digital art online, honestly ask yourself:

- Will this ultimately glorify me or God?
- Will this stir or muffle healthy affections for Christ?
- Will this merely document that I know something that others don't?
- Will this misrepresent me or is it authentic?
- Will this potentially breed jealousy in others?
- Will this fortify unity or stir up unnecessary division?
- Will this build up or tear down?
- Will this heap guilt or relieve it?
- Will this fuel lust for sin or warn against it?
- Will this overpromise and instill false hopes in others?

In asking these questions, I don't want to become so paralyzed with fear that I don't share, and I don't want to be so naive that I become negligent in what I do share. As an online creator, I need these questions to interrogate my heart every time I post or publish.

In all of this, I do not dismiss the value of chitchat or humorous self-deprecation online. These can be powerful tools for missional purposes. Even the apostle Paul did not hesitate to self-deprecate.[41] It didn't paralyze him; it grounded and personalized him as a saved sinner. There is a strategic self-deprecation that comes through laughing at ourselves and that makes our platform more approachable by others. The more God uses you online, and the more you build relationships online with people who do not know you personally, the more persuasively you can use humor to humanize yourself and even to make your message of grace more poignant. For Christians, humor is not an end in itself, but a means of ultimately making gospel truth more real to people watching you online (as we will see later).

Humor or not, self-expression alone is never an adequate reason for Christians to communicate online. To what eternal destiny am I influencing others, and even myself? With this high calling in mind, Paul pleads for prayer. Each of us must know when to speak as we pray, like Paul, that "words may be given to me in opening my mouth boldly to proclaim the mystery of the gospel" (Eph. 6:19).

And we must know when to be silent, too. The virtues of our age are hyperconnectivity and multitasking, not solitude and meditation. But true wisdom calls for word restraint.[42] The same gospel that gives us words to speak teaches us what *not* to say.[43]

In the gospel, we find our message and our commission in the digital age. So we pray, "Lord, let no corrupting talk come off my thumbs, but only what is good for building up, as fits the occasion, so that my social-media investment will give grace to those who see it."[44]

---

41. 1 Cor. 15:9; 1 Tim. 1:15.
42. Prov. 10:19; 11:12; 12:23; 13:3; 15:28; 17:27–28; 18:13; 21:23; 29:20.
43. Titus 3:1–11.
44. See Eph. 4:29.

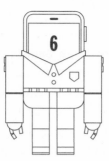

WE BECOME LIKE WHAT WE "LIKE"

The words and images we *share* on our phones influence others (as we saw in the last chapter). But the words and images we *consume* transform us.

Do you remember the story of Narcissus? He was an attractive chap, but he was also arrogant and incapable of receiving love or giving love to anyone. For his frigid affection, the goddess Nemesis cursed him in a most hopeless way, making him fall in love with the image he projected of himself. Day after day, he bent over and caught his reflection in the glassy surface of the water, longing for the image he saw, so much so that one day he noticed his reflection in the bottom of a well, jumped in, and drowned.

It's awkward to say it this way, but like Narcissus staring down into the water, enchanted with himself, we bend over our phones—and what most quickly captures our attention is our own reflection: our replicated images, our tabulations of approval, and our accumulated "likes." Social media has become the new PR firm of the brand Self, and we check our feeds compulsively and find it nearly impossible to turn away from looking at—and loving—our "second self."[1]

---

1. Sherry Turkle, *The Second Self: Computers and the Human Spirit* (Cambridge, MA: MIT Press, 2005).

So when we talk about "smartphone addiction," often what we are talking about is the addiction of looking at ourselves.

FITTING IN

Digital narcissism—this constant, bent-over focus on our own reflections—cannot define our identity in a satisfying way, and there are many reasons why. Fundamentally, finding our identity is not just a matter of self-love but also of conformity.

We know teenagers strive desperately to fit in, and we know that in search of this conformity, they try to stand out. For example, a teen may present himself or herself with jet-black hair, dark eyeliner, and a black wardrobe. This fashion may be an attempt to stand out, but more important, it is an attempt to fit in (to the goth subculture).

But we all do this: we all wear "costumes" to meet the approval of certain subcultures, because our search for individuality is always a chase for conformity. There is an old adage that says, "We are not who we think we are; we are not even who others think we are; we are who we *think* others think we are." In other words, what we think others think of us profoundly shapes our sense of identity and our search for belonging. This complex social dynamic further proves that we don't find our identity in ourselves.

Long before the smartphone, pastor Tim Keller explained this dynamic to his urban congregation. "People in New York City like to think, 'We're individuals. People here can decide what they want to be and do it.' That's not true," he corrected. "You all have your uniforms. Some of you are wearing Wall Street uniforms. Some of you are wearing East Village uniforms. Some of you are wearing SoHo uniforms. There are uniforms! You have to fit in. You have to get your validation from somebody. You have to have a group of people that say, 'You're one of us.'"[2] At the core of our lives, we want to fit in to find our identity.

---

2. Timothy Keller, sermon, "Built Together; Redeemer's Organization Service," Gospel in Life, gospelinlife.com (June 2, 1991).

LIKE MIKE

As we seek to belong, we are brought back to celebrities, too. They provide astoundingly potent models for our collective awe and emulation. In fact, the craving for conformity explains the commercial value of celebrities, and one of sports marketing's greatest feats was accomplished through brute honesty about this conformity. Gatorade's 1992 ad campaign around basketball great Michael Jordan was simple: "Like Mike, I want to be like Mike." Who doesn't want to wear Mike's sneakers, adopt the swagger of Mike, and play ball like Mike? Millions of young athletes want to be like Jordan, so they attempt to emulate his basketball skills even to this day with the refrain, "Like Mike, I want to be like Mike."

By 2016, a new tagline had emerged for Jordan's Nike shoe line: "I am not Michael, I am Jordan"—a brilliant attempt to make more space for individuality under the umbrella of community conformity. Now, many years after his retirement from the hardwood, a lot of people still want Mike's sneakers, and the former basketball star cashes in $100 million a year from his shoe line.

And for those celebrities and athletes still in their prime, emulation pays big, because they represent the glory we want to possess ourselves. To behold majesty is a phenomenon that begins to chip and sculpt the contours of our identity. The desire to imitate the glory we see in others is one of the most obvious (and most profound) psychological realities that advertising targets. We crave acceptance, and we are always becoming like what we admire. So in whose identity will I find my home?

CHANGED BY LOVES

We are composites of the people we want to conform to, and this conformity defines one of the most powerful lures of our smartphones. Digital technology now accelerates and particularizes our search for belonging.

To help explain this phenomenon, I contacted theologian Richard Lints, who has studied how we become *like* what we worship. He examines our conformity in the contexts of both the negative (idolatry) and the positive (worship and sanctification). "We are mirrors," he told me. "And so the whole metaphor of the human being—reflecting its environment, reflecting its context, reflecting its idols, reflecting its gods—is absolutely core, from the beginning to the end of the canon [of Scripture]. What we call worship—worshiping God faithfully and truly—is also a matter of our identity. That is what we are created for. That is who we are."[3]

Whether or not we see it, worship is the fundamental dynamic of our molding. And this is why, no matter how fiercely independent we are, we never find our identity *within* ourselves. We must always look *outside* of ourselves for identity, to our group fit and to our loves. Both dynamics reveal the truth: we are becoming *like* what we see. We are becoming *like* what we worship. Or, to put this in Facebook terms directly, we are becoming *like* what we *like*.

WORSHIP GUIDED AND MISGUIDED

The Bible sharpens the point of this dynamic like a woodworker's chisel. Either we worship what is created (idols) or we worship the Creator (Christ). These are our only options.

If we worship idols, we become like the idols.[4] Idolatry is the vain attempt to find ultimate meaning in finite things that we can craft and hold in our hands. This is extremely clear in Scripture: to love and worship a dead idol is to become like the idol. If our idols have no hands to embrace us, no eyes to see us, no mouths to assure us, and no ears to hear us, then we who worship idols become like them: spiritually powerless, blind, mute, and deaf. Our idols dehumanize us; they petrify our souls, and dumb and dull and deaden all of our spiritual senses.[5] Idols can only distort us (as we'll see more fully later). Therefore, to worship anything that is not God is to fundamentally live in identity confusion.

---

3. Richard Lints, interview with the author, "Why We Never Find Our Identity Inside of Ourselves," Desiring God, desiringGod.org (Aug. 31, 2015).

4. Rom. 1:18–27.

5. Pss. 115:4–8; 135:15–18.

When we worship the glory of our celebrities (like Mike), they become idols of our admiration and conformity, raised up for human "adoration, veneration, and beatification, in the expression of a properly religious sentiment."[6] The age of the spectacle produces the celebrities who become the cultural idols of worship and emulation. But while they may perform for us, and we may adore them like fangirls, our idols do not love us back. They'll never see us.

If we worship Christ, we become like Christ.[7] Opposite our idols, to love and to worship Christ is to become like him, powerfully conforming to his beautiful image, the true image of God. Jesus Christ is the full image of what you and I were created to express.[8] I am made in his image. But my humanity is sinful, twisted, and broken. He loved me so much he shed his blood for me, in order to free me from all other conformity traps.[9] In him, I have been made spiritually alive and given eternal hope and lasting joy, and in him I find the anticipation of a moment when I will see him face to face and experience the full and perfect recovery of everything I was created to be as God's image bearer. This hope and longing is what drives me to see him in Scripture—and then to love him, to reflect him, and to conform to his life now (and anticipate becoming fully conformed to his image in the resurrection).[10]

The object of our *worship* is the object of our *imitation*. God designed this inseparable pattern. What we want to become, we worship. And what we worship shapes our becoming. This is Anthropology 101.

MADE IN GOD'S IMAGE

But all of this talk of mirrors and idols and conformity has not exactly answered the pinnacle question of our identity: Why do I exist?

Of course, we will not find our life's purpose lurking in our

---

6. Jacques Ellul, *The Technological Bluff* (Grand Rapids, MI: Eerdmans, 1990), 382.

7. Rom. 12:1–2; 2 Cor. 3:18; Col. 3:10.

8. 2 Cor. 4:4.

9. Rom. 5:8.

10. 2 Cor. 3:18; 1 John 3:2–3; 1 Cor. 15:42–49.

social-media validation.[11] For the answer, we turn to the Bible, and there we read that we were created by God to image God.[12] To image God means many things—spiritually, rationally, emotionally—but to get to the essence of image bearing, I asked John Piper to explain it, which he did with marble statues: "You put up a statue of Stalin because you want people to look at Stalin and think about Stalin. You put up a statue of George Washington to be reminded of the founding fathers. Images are made to image." What does this mean for flesh and blood? It means God "created little images of himself so that they would talk and act and feel in a way that reveals the way God is. So people would look at the way you behave, look at the way you think, look at the way you feel, and say, 'God must be great, God must be real.' That is why you exist."[13] In other words, we were created to stand in opposition to the techno-worldliness that inevitably makes God look irrelevant in the new world of technique and device mastery.

Here's the key: "God didn't create you as an end in yourself. *He is the end; you are the means.* And the reason that's such good news is because the best way to show that God is infinitely valuable is to be supremely happy in him. If God's people are bored with God, they are really bad images. God is not unhappy about himself. He is infinitely excited about his own glory."[14]

To be made in God's image means we exist for two reasons: (1) to be satisfied in the infinite worth of the Creator and (2) to show the world how precious and deeply satisfying he is. Our "fit," our "loves," and our "belonging" all converge in him. Our identity hinges on him, and in him we find the Spirit-given power

11. Katie Couric: "We spend so much time these days, I think, looking for external validation—with our carefully crafted Instagrams, clever postings, perfect pictures, counting our likes, favorites, followers and friends—that it's easy to avoid the big questions: Who am I? Am I doing the right thing? Am I the kind of person I want to be?" "Katie Couric to Grads: Get Yourself Noticed," *Time* magazine (May 18, 2015).

12. Gen. 1:26–27; 5:1; 9:6; James 3:9.

13. John Piper, interview with the author via Skype, published as "What Does It Mean to Be Made in God's Image?" Desiring God, desiringGod.org (Aug. 19, 2013).

14. John Piper, sermon, "The Story of His Glory," Desiring God, desiringGod.org (Sept. 10, 2008), emphasis added.

to reject all identities projected on us.[15] But if people see us bored with God, absorbed with ourselves, and conformed to worldly celebrities, they will not see the image of Jesus reflected in us. If we fail to reflect Christ, we fail to be what God created us to be; we lose our purpose.

DEVICE WORSHIP

This brings us back to our phones. Our worship and our idolatry are always acts of surrender, writes Peter Leithart on our tendency to yield ourselves to our technology: "Idolaters of technology don't literally consider their technologies to be divine. But many do 'lower' themselves before their technologies. Instead of wisely using the products of their labor and ingenuity, they 'bow' until the latest gimmick is ruling their lives—determining how they use their time, how they spend their money, their interests and values."[16] Submission to a created thing, such as a smartphone, is idolatry when that created tool or device determines the ends of our lives.

This form of idolatry—submitting human *ends* to the available technological *means*—is called *reverse adaptation*.[17] In the digital age, we idolize our phones when we lose the ability to ask if they help us (or hurt us) in reaching our spiritual goals. We grow so fascinated with technological glitz that we become captive to the wonderful *means* of our phones—their speed, organization, and efficiency—and these means themselves become sufficient *ends*. Our destination remains foggy because we are fixated on the speed of our travel. We mistakenly submit human and spiritual *goals* to our technological *possibilities*. This is reverse adaptation.

Our idolatrous impulses make us easily trapped by this worldliness,

---

15. Rom. 12:2.

16. Peter J. Leithart, "Techno-god," *First Things*, firstthings.com (Sept. 27, 2012). "By continually embracing technologies, we relate ourselves to them as servo-mechanisms. That is why we must, to use them at all, serve these objects, these extensions of ourselves, as gods or minor religions." Marshall McLuhan, *Understanding Media: The Extensions of Man* (Cambridge, MA: The MIT Press, 1994), 46.

17. Langdon Winner, *Autonomous Technology: Technics-out-of-Control as a Theme in Political Thought* (Cambridge, MA: MIT Press: 1977), 229.

the loss of our purpose. We often don't stand over our phones and direct them, based on our calling to image God; instead, we bow to our phones as worlds of digital possibilities, never asking the questions of our ultimate aims. When the *means* become our aimless habits, this is techno-idolatry.

THE IDOLS OF SOCIAL MEDIA

If idols shape us, unhealthy phone patterns are bound to be reflected in our relationships.

Our digital interactions with one another, which are often necessarily brief and superficial, begin to pattern all our relationships. When our relationships are shallow online, our relationships become shallow offline. Douglas Groothuis, a professor of philosophy at Denver Seminary, warns: "The way we interact online becomes the norm for how we interact offline. Facebook and Twitter communications are pretty short, clipped, and rapid. And that is not a way to have a good conversation with someone. Moreover, a good conversation involves listening and timing, and that is pretty much taken away with Internet communications, because you are not there with the person. So someone could send you a message and you could ignore it, or someone could send you a message and you could get to it two hours later. But if you are in real time in a real place with real bodies and a real voice, that is a very different dynamic. You shouldn't treat another person the way you interact with Twitter."[18] Yet our online habits change our relational habits: both become clipped and superficial, and we become more easily distracted and less patient with one another.

Our relationships also suffer when our thinking becomes caught in the ebb and flow of online fiascos. Writer Alan Jacobs spent seven years on his iPhone, seven years engaged on Twitter, and more than ten years responding to blog comments. Then he stood back, evaluated it, and dropped it all. He ditched social

---

18. Douglas Groothuis, interview with the author via phone (July 3, 2014).

media and his iPhone.[19] "I have considered the costs and benefits," he said, "and I have firmly decided that I'm not going to be held hostage to that stuff anymore." Why not? "The chief reason is not that people are ill-tempered or dim-witted—though Lord knows one of those descriptors is accurate for a distressingly large number of social-media communications—but that so many of them are blown about by every wind of social-media doctrine, their attention swamped by the tsunamis of the moment, their wills captive to the felt need to respond now to what everyone else is responding to now."[20]

When Andrew Sherwood, a graduate student, decided to do the same (ditch social media and the smartphone), his wife said it was the greatest gift he ever gave her. Why? "When you had your smartphone, you were a walking vending machine of whatever you'd ingested that day," she told him. "It was difficult to talk about deeper things that mattered, because you were constantly distracted by Internet litter. You're now able to focus and give necessary attention to deeper issues. More of what we talk about comes from your heart rather than your Twitter feed."[21] Whether or not it's time to ditch your smartphone altogether is a question we will save for later, but Andrew offers us a graphic illustration of how digital idols pattern us.

THE WARNING AND THE HOPE

As human beings, we were made to image God, which means our identity is, by definition, moldable, and that means susceptible. We are statues of wax, changed and reshaped by what we do on our phones. But this pliability also means we can be redeemed, remade, and restored by the sovereign grace of our image-sculpting Savior to do what we were made to do: magnify God. As we image him, we

---

19. Alan Jacobs, "My Year in Tech," *Snakes and Ladders*, blog.ayjay.org (Dec. 23, 2015).

20. Alan Jacobs, "I'm Thinking It Over," *The American Conservative*, theamericanconservative. com (Jan. 4, 2016).

21. Andrew Sherwood, "The Sweet Freedom of Ditching My Smartphone," *All Things for Good*, garrettkell.com (Jan. 21, 2016).

invite the world to a welcoming Father, where the lost can find refuge and identity, and where thirsty sinners will find the all-satisfying living water.

True image bearing frees us to be digitally honest about ourselves. We pray for grace to avoid the plight of Narcissus—to avoid falling in love with the image of ourselves. And we pray for grace and courage to take a more honest look at our digital reflections in the glossy screens of our phones and see where we fail to image Christ, willing to humbly admit, repent, and change when we sometimes see the reflection of a dragon looking back.

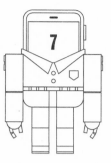

WE GET LONELY

A middle-aged homeless man sits alone on a sunny city sidewalk, back against a fence, dozing. Karim, a generous passerby, approaches and stands above him with cash in hand. The man on the street startles awake, flinches in self-defense, and clutches his backpack of belongings. As his eyes adjust to the sunlight, he sees the outstretched hand and takes the money with gratitude.

They begin to chat, and the homeless man introduces himself as Mark. In a role reversal, Mark grabs his grubby backpack, asks Karim to wait a moment, stands up, and walks off with the cash, leaving Karim alone on the street. Mark returns moments later with a plastic bag and two Styrofoam boxes. Mark used the handout to buy two dinners—one to share.

"Please sit and eat with me for a little bit?" Mark asks.

Karim is surprised, but agrees and sits down on the concrete.

"I'm glad you're here with me," the homeless man says, as they sit on the sidewalk and unbox their dinners together. "It's lonely out here. People walk by and they ignore me. They could care less if I was dead or alive. It's great just to sit out here with somebody."

Video of the exchange was captured through a hidden camera, and every time I see it on YouTube, my heart is stirred.[1]

---

1. Karim Metwaly, video, "Lonely Homeless Man," YouTube, youtube.com (June 19, 2015).

ONE THOUSAND SHARDS OF GLASS

One never knows the authenticity of videos such as this, but this particular one spread virally, and it's easy to see why. The video exposes a side to homelessness that is mostly ignored and rarely captured. More fundamental to human life than money, food, and shelter is human friendship. We were made to connect with other humans for true fellowship, all because we were made in the image of the triune God. And this is why loneliness stings like an open gash in our skin.

J. H. van den Berg, the late Dutch psychiatrist, famously wrote, "Loneliness is the nucleus of psychiatry." He also wrote, "If loneliness didn't exist, we could reasonably assume that psychiatric illnesses would not occur either."[2] To these stunning quotes, theologian Peter Leithart adds this spiritual interpretation: "Humans connect to other humans at so basic a level that when we disconnect, our souls shatter into a thousand little pieces."[3]

I think I can understand the link between loneliness and homelessness. What is more difficult to understand is why such rampant loneliness persists in the hyperconnected digital age.

ONLINE AND LONELY

Smartphones and social media were supposed to cure the epidemic of loneliness. We would all be connected—all together, all the time—and none of us would ever feel alone. But the harsh truth is that we can always be lonely, even in a crowd—and now, even more so, in a digital crowd.[4]

We send texts, pictures, and videos; we post tweets and Facebook updates; and we refresh and wait—often looking to a stagnant screen that shows no responses, or very few. When we hit refresh and stare at a screen with no new updates, it can seem that no one is on the other

---

2. Cited in Peter J. Leithart, *Traces of the Trinity: Signs of God in Creation and Human Experience* (Grand Rapids, MI: Brazos, 2015), 17.

3. Ibid.

4. Katie Couric: "Social media can be a great thing: giving voice to the voiceless, uniting people across the globe in a common cause. But proceed with caution. Constant connectivity can leave you feeling isolated and disconnected. Do not be seduced by the false intimacy of social media." "Katie Couric to Grads: Get Yourself Noticed," *Time* magazine (May 18, 2015).

side. We feel the sting of loneliness in the middle of online connect-edness. Sometimes we feel as if we are walking through a museum of relational relics and holograms. In reality, "it's a lonely business, wandering the labyrinths of our friends' and pseudo-friends' pro-jected identities, trying to figure out what part of ourselves we ought to project, who will listen, and what they will hear."[5]

It's a chicken-or-egg question: Does Facebook make us lonely or does it appeal to those of us who are already lonely? That debate is hard to resolve, but it makes one point clear: we have begun giving up on the idea that Facebook, the map of all our human networks, can end our loneliness.

TECHNOLOGY AND ISOLATION

In the big picture, technology offers us many benefits, but with one major pitfall: isolation. Isolation is both the promise and the price of techno-logical advance. "The problem is that we invite loneliness, even though it makes us miserable," writes author Stephen Marche. "The history of our use of technology is a history of isolation desired and achieved."[6]

The long story of isolation desired and achieved is retold by Giles Slade in his book *The Big Disconnect: The Story of Technology and Loneli-ness*.[7] There he shows how many strands of technology and loneliness have been woven together in the history of various innovations, from street peddlers and phones to television and music.

As technology improves, machines replace people and automation replaces interaction. Street vendors gave way to vending machines. Fresh milk deliveries gave way to dairy products preserved in refrig-erators. Bankers gave way to ATMs. Two hundred years ago, laborers were personally acquainted with their clients. In today's technological society, many laborers work in remote locations, in industrial or busi-ness parks, serving faceless clients or nameless consumers from whom they are separated geographically or by a very long production chain.

---

5. Stephen Marche, "Is Facebook Making Us Lonely?" *The Atlantic* magazine (May 2012).
6. Ibid.
7. Giles Slade, *The Big Disconnect: The Story of Technology and Loneliness* (Amherst, NY: Pro-metheus, 2012).

Physically, we are drawn apart by other factors. Gathering around a fire gave way to central heating, which pushes heat to all the corners of the house. Gathering together for the local news at a pub gave way to the reading of newspapers, creating a paper wall shielding our faces from one another.

Isolation was later deepened by advances in video. The community cinema gave way to a large shared television in each family's home, which gave way to portable televisions, and now to personal LED TVs in every bedroom.

When it comes to music, this technological trajectory is even clearer. Attending a live orchestra performance on a Saturday evening was, for many people, replaced by the stationary phonograph (or record player) in the family room, which was replaced by a large transistor radio, which was replaced by a portable transistor radio, which was replaced by a boom box with open speakers carried on the shoulder, which was replaced by a Walkman clipped to the belt, which was replaced by a tiny iPod clipped to the sleeve. Music went from a social community experience to a shared family experience to a private earbud experience.

Technology is always drawing us apart, by design. Our isolation is desired and achieved.

OUR PORTABLE SHIELDS

Many of these technological trajectories converge in the smartphone—the supreme invention of personal isolation. Our smartphones are portable shields we wield in public in order to deter human contact and interaction. When we step into an occupied elevator, we grab our phones like security blankets.

Headphones extend this principle to the ultimate degree. By definition, to lock into our earbuds is to refuse to listen to silence, and "a refusal to listen to silence is a refusal to meet oneself or others."[8] By them, we close ourselves off from the outside world,

---

8. Jacques Ellul, *The Technological Bluff* (Grand Rapids, MI: Eerdmans, 1990), 378.

but we also close ourselves off from ourselves (à la Blaise Pascal). Headphones give us a buffer from both healthy introspection and social conversation.

"In the twenty-first century, glaringly white Apple earbuds inform all those who observe us in public that we are disinterested, musically inclined, non-threatening people, while Bluetooth WiFi earpieces convey a slightly different, more aggressive message: far too busy, don't dare disturb," Slade writes. "Once again, interaction with a device prevents and is preferable to risky, energy-consuming interactions with strangers. We have been conditioned for over a hundred years to risk interpersonal contact only through the mediation of machines. We trust machines much more than we trust human beings."[9] Reminiscent of Andy Warhol's sound recorders and Polaroid cameras, our machines now buffer (and broker) our relationships.

By preserving our isolation, we unwittingly walk right into one of the world's most brilliant marketing traps. "For manufacturers and marketers, human beings are best when they are alone, since individuals are forced to buy one consumer item each, whereas family or community members share," writes Slade. "Technology's movement toward miniaturization serves this end by making personal electronics suitable for individual users. For today's carefully trained consumers, sharing is an intrusion on personal space."[10]

ROLE REVERSING

The miniaturization and personalization of technology, the direction of many of our technological advances, cuts us off from others in much of our ordinary interactions. We seek to exert control over others by mediating our relationships through technology. "In a technicized culture, communal ties are readily cut and replaced by technical or organizational relations. Love dies; empathy and

---

9. Slade, *The Big Disconnect*, 160.
10. Ibid., 10.

sympathy and contact with the other disappear. Estrangement and loneliness increase."[11] That's overstating it, but, as Slade noted above, we do seem to trust people less than we trust our technology.

As if we are having a conversation through glass and wall-mounted telephones at a prison visitation, many of us now approach one another from behind safe barriers, with a digital sign language of taps, swipes, and multitouch gestures on screens. Even when we are with our closest friends and family members, we are drawn back to our online networks. (In the slow moments during vacations or gatherings, how many people can you find on their phones?)

The smartphone is causing a social reversal: the desire to be alone in public and never alone in seclusion. We can be shielded in public and surrounded in isolation, meaning we can escape the awkwardness of human interaction on the street and the boredom of solitude in our homes. Or so we think.

BUILDING FACE-TO-FACE TRUST

On top of all this, the technological age expedites physical dislocation, says theologian Kevin Vanhoozer. "One of the problems with globalization, transportation, and communications technology, and modernity in general, is that these benefits also come with a cost: *displacedness*. The result of our ability to talk to people anywhere in the world instantaneously, or to travel to the other side of the planet in a matter of hours, is a loss of the sense of belonging to any one particular place. Distance is no longer an impediment. That's potentially a good thing, to be sure. But, on the other hand, our connectedness to places near and far makes it harder for any one place to feel like home."[12]

Perhaps more concerning are the relational implications. If we have no hometowns, we are more likely to isolate ourselves and to expect distant relationships to root us. But if our deepest and most treasured relationships are remote, we are brought back to concerns

---

11. Egbert Schuurman, *Faith and Hope in Technology* (Toronto: Clements, 2003), 101.
12. Kevin Vanhoozer, interview with the author via email (Feb. 26, 2016).

about the frictionlessness of our smartphone touch screens and our need for the rough edges of face-to-face interactions. This is where the advantages of embodied awkwardness come into play. The most shaping conversations we need are full of friction, and we simply cannot have them on our frictionless phones.[13]

And when it comes to interacting with strangers, social media emerges as a safe place to do it. Perhaps it's not going too far to say that we love social media "because it comes without the hazards and commitments of a real-world community" or because we really harbor "a deep disappointment with human beings, who are flawed and forgetful, needy and unpredictable, in ways that machines are wired not to be."[14] It is safer to approach one another from behind a machine.

Social media feels like a safe way to offer ourselves to others. On a phone screen, testifies one writer, "I could put myself out for virtual inspection and validation while remaining in control, remote from the possibility of physical rejection."[15] But while we may be able to browse suitors by casually flipping through profiles in a dating app, we know we cannot choose a mate that way. We need face-to-face time, and even then we are hardly prepared for the friction that God intends to use as we and our spouses are sharpened and shaped over the years into couples who reflect Christ and his bride. This is part of the genius (and the mystery) of marriage as a covenant bond between two people of differing genders and often differing ethnicities, talents, and interests.

Online, we offer up our lives in stories forged by self-interpretation, and only rarely is our interpretation called into question. In person, however, our interpretations can be pushed back, questioned, and challenged, all for our own good.

Friction is the path to genuine authenticity, and no amount of

---

13. Prov. 27:17.

14. Jonathan Franzen, "Sherry Turkle's 'Reclaiming Conversation,'" *The New York Times* (Sept. 28, 2015).

15. Olivia Laing, *The Lonely City: Adventures in the Art of Being Alone* (New York: Picador, 2016), 224.

online communication can overcome a lack of real integrity. We must be real with the people God puts into our lives. We must tell the truth. We must be honest at school. We must be wise with our money. We must be trusted friends. We must be reliable at work. The world needs what we must be: God-centered, joyful, and trustworthy men and women. We are not flawless; we are fallen repenters who require relational friction to grow and mature. We are authentic believers who are committed to replacing easy relationships with authentic ones.

From this embodied authenticity, the gospel spreads.[16] Wherever we live, Christians are called to engage the world face to face—a key point for us all, especially parents, to keep in mind. "I meet more and more kids that don't know how to talk to people, and who don't even want to look up from their screens," Francis Chan told me. "We are raising soldiers. We are raising missionaries. Our job is to get these kids to where they can get into the world and start conversations with people and bring the light of Jesus and the message of the gospel to them."[17] Eye-to-eye authenticity is the key to empathy, humility, and trust in our relationships, and these are skills we all need.

PROTECT ALONENESS

At the same time, face-to-face authenticity needs true aloneness.

Sherry Turkle, a respected psychologist of the digital age, says: "The capacity for empathic conversation goes hand in hand with the capacity for solitude. In solitude we find ourselves; we prepare ourselves to come to conversation with something to say that is authentic."[18] Solitude is a precious gift: we all want it, we all need it, and we all think more technology is the secret. It's not, warns Alastair Roberts. "I fear that our hyperkinetic, cacophonous, and riotous audio-visual environments erode the art of silent and attentive listening, and with it, our sense of the presence of the invisible."[19]

---

16. 1 Thess. 1:2–10.
17. Francis Chan, interview with the author, "Dads and Family Leadership," Desiring God, desiringGod.org (Jan. 13, 2015).
18. Sherry Turkle, "Stop Googling. Let's Talk," *The New York Times* (Sept. 26, 2015).
19. Alastair Roberts, interview with the author via email (Jan. 23, 2016).

So what do we do with all the aloneness afforded to us in the technological age? We often fumble it by wrongly using our technologies. You'll remember that in the first chapter I mentioned my survey of eight thousand Christians about their social-media routines.[20] I noted there that more than half of the respondents (54 percent) admitted to checking their smartphones within minutes of waking up on a typical morning. When asked whether they were more likely to check email and social media *before* or *after* having their spiritual disciplines on a typical morning, 73 percent said *before*.

This reality is concerning if John Piper was right when he said: "I feel like I have to get saved every morning. I wake up and the devil is sitting on my face."[21] Those early morning hours are vital for spiritual health and for making progress in the spiritual battles we face every day.[22] Satan knows it, and he wants to destroy our devotional life, and if he cannot get us to simply ignore the habit, he will "distract [our] thoughts, and break them into a thousand vanities."[23]

It is no surprise that we relinquish our morning hours by turning to our phones, but why? What is the lure? I asked Piper, and he pointed to six instinctive reasons—three "candy motives" and three "avoidance motives."

1. *Novelty Candy*. We want to be informed about what is new in the world and new among our friends, and we don't want to be left out of something newsworthy or noteworthy.
2. *Ego Candy*. We want to know what people are saying about us and how they are responding to things we've said and posted.
3. *Entertainment Candy*. We want to feed on what is fascinating, weird, strange, wonderful, shocking, or spellbinding.

---

20. A non-scientific survey of desiringGod.org readers, conducted online via social-media channels (April 2015).

21. Apparently this statement was excerpted from a message, paraphrased, and spread online. The original sermon is unknown. My wording here was confirmed and approved by John Piper via email (June 2, 2015).

22. As was true of the psalmists—Pss. 5:3; 88:13; 90:14; 119:147–48; 130:6; 143:8.

23. John Flavel, *The Whole Works of the Reverend John Flavel* (London: W. Baynes and Son, 1820), 4:253.

4. *Boredom Avoidance.* We want to put off the day ahead, especially when it looks boring and routine, and holds nothing of fascination to capture our interest.
5. *Responsibility Avoidance.* We want to put off the burdens of the roles God has given us as fathers, mothers, bosses, employees, and students.
6. *Hardship Avoidance.* We want to put off dealing with relational conflicts or the pain, disease, and disabilities in our bodies.[24]

Perhaps we check our phones for more noble ends—to communicate with friends and family members or to confirm our schedules for the day—but a rush of temptations comes at us immediately in the morning, and we fumble our precious solitude. It's hard to summarize the resulting problem any better than this: "The real danger with Facebook is not that it allows us to isolate ourselves, but that by mixing our appetite for isolation with our vanity, it threatens to alter the very nature of solitude."[25] These equations seem to hold true for our early morning hours:

Isolation + feeding on vanity = soul-starving loneliness

Isolation + communion with God = soul-feeding solitude

The bottom line: technology bends us in a centripetal direction, pulling us toward a central habitat of loneliness and filling our lives with habits that benefit the stakeholders who seek to monetize our attention.

And when it comes to the morning hours, Charles Spurgeon was right: "Permit not your minds to be easily distracted, or you will often have your devotion destroyed."[26] Vital to our spiritual health, we must listen and hear God's voice saying to us, "Be still, and know that I am God" (Ps. 46:10). Every morning we must take time to stop, to be

---

24. Summary of John Piper, interview with the author via Skype, published as "Six Wrong Reasons to Check Your Phone in the Morning: And a Better Way Forward," Desiring God, desiringGod.org (June 6, 2015).
25. Marche, "Is Facebook Making Us Lonely?"
26. C. H. Spurgeon, *The Sword and Trowel: 1878* (London: Passmore & Alabaster, 1878), 136.

still, to know that God is God and that we are his children. Digital technology must not fill up all the silent gaps of life.

So as Christians, we push back our phones in the morning—in order to protect our solitude so that we can know God and so that we can reflect him as his children. And we push back our phones during the day—in order to build authentic eye-to-eye trust with the people in our lives and in order to be sharpened by hard relationships. Without these two guards in place, our displacedness dominates, isolation shelters us, we can find ourselves becoming more and more lonely, and our gospel mission will eventually stall out.

But there remains an even more lurid smartphone habit that thrives under the veil of secrecy.

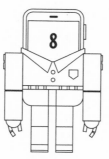

WE GET COMFORTABLE IN SECRET VICES

By nature, we are needy consumers. We are designed to take and eat, to receive material gifts in order to survive, and to drink from the water of life.

However, *consumerism* is the idea that all of life can be converted into commodities, then controlled and monetized. The catchphrase "There's an app for that" is a reigning motto for the consumerist spirit in the smartphone age. Today, all of our activities and interests (and even our relationships) can be rendered into discrete tabulations, like in Snapchat, where relational connections are reduced to points and where "Snapstreaks" can be maintained by connecting at least once every twenty-four hours with particular friends. In fact, "the technology of social media is becoming more 'gamified' by the year as developers learn how to tap into the deep human hunger for simulations of authority and vulnerability."[1]

In a culture that can reduce relationships to a personal score in a competitive game, every experience, hope, and longing in life can just as easily be rendered into digital merchandise—even the most intimate parts.

---

1. Andy Crouch, *Strong and Weak: Embracing a Life of Love, Risk and True Flourishing* (Downers Grove, IL: InterVarsity Press, 2016), 86.

ASHLEY MADISON

Ashley Madison is a Canadian web-based subscription service targeting married men and women seeking to initiate anonymous connections with other aspiring adulterers. The site's slogan could not be more simple (or insipid): "Life is short. Have an affair." The site did with sex and relationships what digital technology tries to do with all of life—it made them consumable commodities. It turned adultery into a commodity that, for a fee, users could acquire by discreetly submitting their email addresses to a database and becoming members who could message other members to coordinate secret adulterous rendezvous. Over the years, tens of millions of people secretly registered their names, credit cards, email addresses, and home addresses, and even wrote out their sexual fantasies.

But over time, many users apparently had second thoughts after registering and went back into the site to delete their accounts and personal information. Whether or not the profiles and registration information were actually deleted from the company servers was a cause of dispute, so to find out, a team of hackers broke into the site in the summer of 2015 and discovered that no information had ever been permanently removed from the database. The hackers then stole all the data records and leaked the names and email addresses publicly.

News of the data breach stirred waves of fear. Suspicious spouses everywhere took the next dreaded step of searching the online databases to see if their own beloveds' names or email addresses were included.

Samantha, an alias for one such forty-eight-year-old wife, recounts finding her husband's email address among the leaked data while she was at work. Stunned, she grabbed her purse and keys, and drove home immediately.

> My husband was in the kitchen and he was surprised to see me home. He knew that something was wrong.
>
> I said, "Look at the pain and the grief on my face, do you see it?"

He said, "I do. What's going on?"

I said, "I found your name and e-mail address on Ashley Madison."

He said, "No, you didn't."

I said, "You know exactly what I am talking about."

He was going pale. He kept swallowing. I know my husband very well: he was in a panic.[2]

Panic, a knot in the throat, a hole in the soul—this was the feeling of millions who could not explain the dark intentions of their hearts as just "slip ups" or mistaken clicks. Their intentions were now exposed to the world and to any loved ones with a hint of suspicion. In one data leak, thirty-two million adulterers (or aspiring adulterers) were outed, including military personnel, prominent celebrities, and even pastors and ministry leaders. Suicides followed (including one fifty-six-year-old pastor and seminary professor).

Most tragic, it now appears that Ashley Madison was really just a gigantic scam targeting naive men. Investigations showed that of the thirty-two million profiles, only twelve thousand were active accounts of real women. When the data from the leaks was further studied, a tabulation was made of users who were actively checking their message inboxes. The breakdown was 20.3 million men to fifteen hundred women, a ratio of more than thirteen thousand men for every one woman. "When you look at the evidence, it's hard to deny that the overwhelming majority of men using Ashley Madison weren't having affairs. They were paying for a fantasy."[3] It was a lie, and millions of men entertained the daydream under the false cloak of anonymity, then got a dagger of reality to the gut.

Technology does this—it makes us think we can indulge in anonymous vices, even conceptually, without any future consequences. Anonymity is where sin flourishes, and anonymity is the most

---

2. Kristen Brown, "I Found My Husband in the Ashley Madison Leak," *Fusion*, fusion.net (Aug. 21, 2015).

3. Annalee Newitz, "Almost None of the Women in the Ashley Madison Database Ever Used the Site," *Gizmodo*, gizmodo.com (Aug. 26, 2015).

pervasive lie of the digital age. The clicks of our fingertips reveal the dark motives of our hearts, and every sin—every double-tap and every click—will be accounted for.

THE PRICE OF CHEAP CURIOSITIES

Tragedies such as the Ashley Madison data breach are heartbreaking, yet they are revealing: they show us that deceptional living in the digital age is convenient. The walls of inconvenience that made vices difficult to act on in previous generations have been lowered or eliminated in the digital age.

First, as indicated earlier, smartphones make sexual sin more discreet, giving it space to fester behind a veil of privacy. Hackers aside, illicit affairs can now be coordinated with a degree of anonymity and secrecy hardly imaginable before smartphones. For singles, when hookup culture meets dating apps, easy sex becomes a widely available commodity. "Flirt apps" such as Tinder use GPS location technology; with little more than a browse of profiles of people nearby and available, a man can swipe right on a woman's image to "tell" her she's attractive. If she replies, the two can open a dialogue and potentially meet in person. As dating apps become more simplified, more visually based, and more geographically fixed, they feed the hookup culture for casual sex and perhaps confuse what men and women are searching for in the app to begin with.[4]

Second, smartphones make free pornography easier to find than the weather forecast. Porn has always been the main driver in visual digital communications, and it is a pervasive problem. In my survey of eight thousand Christians, I found that ongoing porn use is a major issue facing professing believers, mostly young men, although no demographic is immune.[5] More than 15 percent of Christian men over age sixty admitted to ongoing pornography use; the rate was

---

4. Users looking for easy sex and users looking for new relationships naturally meet through the same app, but with widely differing expectations. For more, see Tony Reinke, "Tough and Tinder: Does Easy Sex Make Rude Men?" Desiring God, desiringGod.org (March 12, 2016).

5. A non-scientific survey of desiringGod.org readers, conducted online via social-media channels (April 2015).

more than 20 percent for men in their fifties, 25 percent for men in their forties, and 30 percent for men in their thirties. But nearly 50 percent of professing Christian men ages eighteen to twenty-nine willingly acknowledged ongoing porn use. The survey found a similar trend among women, but in lesser proportions: 10 percent of females ages eighteen to twenty-nine; 5 percent of those in their thirties; and increasingly smaller percentages for those in their forties, fifties, and sixties or beyond. On the one hand, free porn, accessed on a smartphone, is now culturally "a public hazard of unprecedented seriousness."[6] But even more concerning, among Christians, free porn accessed on a smartphone represents a spiritual epidemic of unprecedented gravity in the history of the church, costing a whole generation of young Christians their joy in Christ and corroding young souls by the acid of unchecked lust.

Third, smartphone vices capitalize on our endless curiosities. Under-eighteen pregnancy rates have plummeted in England and Wales since the introduction of smartphones and social media, and no one really knows why—though some researchers suggest that the correlation cannot be explained by new access to contraception or a sudden change in public sex education.[7] A similar cultural phenomenon has been noticed in Japan.[8] It is suggested, among many other factors, that perhaps the curiosities that drove teens to experiment with sex in previous generations are now pacified by sexting and online porn.

Brad Littlejohn explored this dynamic in a 2016 lecture. "Rather than stoking the flames of lust to create testosterone-driven sex monsters, pornography seems if anything to emasculate its users, rendering them passive and impotent," he said. "And I mean 'impotent' here in a clinical as well as a metaphorical sense; no symptom of compulsive pornography use seems to be so widespread as complaints

---

6. Rabbi Shmuley Boteach and former porn star Pamela Anderson, "Take the Pledge: No More Indulging Porn," *The Wall Street Journal* (Aug. 31, 2016).

7. John Bingham, "How Teenage Pregnancy Collapsed After Birth of Social Media," *The Telegraph* (March 9, 2016).

8. Abigail Haworth, "Why Have Young People in Japan Stopped Having Sex?" *The Guardian* (Oct. 20, 2013).

of erectile dysfunction and other sexual disorders. Many porn addicts seem to remain virgins far longer than their peers, struggling to form meaningful relationships with the opposite sex or develop much enthusiasm for sexual activity." In the end, digitally available porn "is driven primarily by that trademark of curiosity, the thirst for novelty, in which the gaze objectifies and devours its object almost immediately, and must move restlessly on to the next, never satisfied."[9]

In an insatiable pornified generation, millions of young men are losing their capacity for human intimacy as they give themselves willingly to this bondage. At the tap of a finger, at any time or place, a beautiful woman will strip off her clothes for you and engage in any lurid sexual act you request, and such easy satisfaction of lust bypasses the rather necessary difficulties that shape a healthy marriage. The ultimate cost of free porn on future marriages is enormous.

Fourth, if curiosity is the impulse driving us to find, watch, and read what is lurid on our phones, perhaps we are witnessing an ancient impulse play out in the digital world. At the creation, God prohibited Adam and Eve from one tree, calling them to self-limit what they wanted to know and experience. They failed in their self-restraint and forced their way to forbidden knowledge. This sin—seeking to satisfy forbidden curiosity—is the hallmark transgression behind all others, and it is no less bold in a consumer-driven economy. We scoff at self-limited understanding of this fallen world, and yet God has said some knowledge is forbidden, because some knowing will destroy us—as seen in the insatiable curiosity that leads into deeper and deeper addiction to more and more lurid forms of pornography. Smartphones make it possible for users to help themselves to fresh forbidden fruit at any moment of any day, and thereby destroy themselves in secret.

GOD SEES ALL

Digital pornography is catastrophic to our souls, not only because it degrades its users, but also because (just like the Mirror of Erised)

---

9. W. Bradford Littlejohn, lecture manuscript, "The Vice of Curiosity in a Digital Age," The Society of Christian Ethics, scethics.org (Jan. 9, 2016).

it exposes the invisible curiosities, idols, and desires of our hearts. Thus, we come to see what God has seen all along.

We fool ourselves with anonymity. But whether it's an abundance of shoes, dirty humor, discreet sexting, illicit pornography, or anonymous adultery, no addiction in our lives is hidden from the eyes of God. Our Creator is no respecter of privacy laws. His omnipresence shatters the mirage of anonymity that drives so many people to turn to their phones and assume they can sin and indulge without consequence.

Not that we are completely unconcerned about consequences: we simply fear the wrong ones. We want to control what information is put online, but our inability to control our online presence leads to personal insecurity. One of the things we hate most is finding unflattering pictures of ourselves posted online by others. And there are wise security issues involved in protecting certain facts about our lives. But fears about what can be found out about us online can also manifest themselves in attempts to shroud private behaviors that are sinful, as in the case of Ashley Madison.

Pornography is the web's largest industry, and the medium fits the vice. But the sobering fact is that our private sexual practices measure our proximity to God.[10] So the stakes could not be higher when it comes to what we do with online allurements, even pervasive and free ones.

And yet, every man who secretly gazes at a nude porn actress has already committed adultery in his heart. So if your right eye causes you to sin, tear it out and throw it away. It is better to lose one eye than have your whole body thrown into hell. And if your scrolling hand causes you to sin, cut it off and throw it away. It is better to lose the capacity to scroll for pornography than have your whole body thrown into hell.[11] In the warning of Sinclair Ferguson, "It is better to enter heaven having decided to never use the Internet again, rather than going to hell clicking on everything you desire."[12]

---

10. 1 Thess. 4:3–5.
11. Matt. 5:27–30.
12. Sinclair Ferguson, interview with the author via phone (Sept. 15, 2016).

Only Scripture tells us what's ultimately at stake here. Data breaches by hackers, shocked discoveries by wounded spouses, and even the self-murder of aspiring adulterers—each of these tragic fallouts of secret sin serves as a mere prophetic hint of an impending reckoning. One day, every sinner who lived in so-called "anonymous" sin will stand before God. There is no such thing as anonymity. It is only a matter of time. Every lurid detail, sleazy fantasy, lazy word, and idle click will be broadcast in the court of the Creator. All of the things done in secrecy and darkness will be brought into the light, and every intent of the heart will be disclosed.[13] It will be the ultimate humiliation. It will be the ultimate exposing of our hearts' intents. It will spark the ultimate panic. It will cause the ultimate knot in the soul and the ultimate desire to run and hide and die from the guilt and shame of being exposed.

Every attempt to bleach-wash our digital footprint is vain. You can delete the most immature images from your Twitter, Instagram, and Facebook feeds. But nothing you do on your phone, have ever done on your phone, or ever will do on your phone is secret. Eternal regret will follow forever for private smartphone clicks happening right now. Before God, our browsing history remains a permanent record of our sin and shame—unless he shows mercy. Before his omniscient eyes, our browsing history can be washed clean only with the blood of Christ.[14]

NOT BY SIGHT

Digital consumerism is directly at odds with many of the most fundamental convictions of the gospel. Spiritual authenticity is measured by faith in the unseen truth of God, not by confidence in the visible consumables of our age. The great term "by faith" is a synonym for confidence in the unseen spiritual realities.[15] Yet on what your heart loves, your eyes will linger.[16] This was true before the photographic

---

13. 1 Cor. 4:5.
14. Col. 2:13–15.
15. Hebrews 11.
16. This is a recurring theme in the book of Isaiah, where the verb "look to" is simultaneously applied to physical sight and spiritual sight (loyalty) in contrasting the categories of idols/God, visual/faith, and the immediate/anticipated.

revolution and before the video revolution. Long before the emergence of digital cameras measured by megapixels and smartphone screens measured in gigapixels, Scripture was vigilant to focus our attention on things unseen.

- "If then you have been raised with Christ, seek the things that are above, where Christ is, seated at the right hand of God. Set your minds on things that are above, not on things that are on earth" (Col. 3:1–2).
- "We look not to the things that are seen but to the things that are unseen. For the things that are seen are transient, but the things that are unseen are eternal" (2 Cor. 4:18).
- "We walk by faith, not by sight" (2 Cor. 5:7).
- "For in this hope we were saved. Now hope that is seen is not hope. For who hopes for what he sees? But if we hope for what we do not see, we wait for it with patience" (Rom. 8:24–25).
- "Now faith is the assurance of things hoped for, the conviction of things not seen" (Heb. 11:1).
- "Jesus said to him, 'Have you believed because you have seen me? Blessed are those who have not seen and yet have believed'" (John 20:29).
- "Though you have not seen him [Christ], you love him. Though you do not now see him, you believe in him and rejoice with joy that is inexpressible and filled with glory, obtaining the outcome of your faith, the salvation of your souls" (1 Pet. 1:8–9).
- "Therefore, preparing your minds for action, and being sober-minded, set your hope fully on the grace that will be brought to you at the revelation of Jesus Christ" (1 Pet. 1:13).
- "For all that is in the world—the desires of the flesh and the desires of the eyes and pride of life—is not from the Father but is from the world. And the world is passing away along with its desires, but whoever does the will of God abides forever" (1 John 2:16–17).

Ignore these passages, and the Christian life makes no sense.

To be sure, we must not think of this as a simple trade—the invisible rendering the visible pointless. No. Rather, as faithful eyes perceive the unseen glories of God and reborn hearts embrace them, all the visible glories of God in the world seem to thicken in substance. The more eagerly we embrace God, the more gratitude we express for his created gifts to us[17] and the more clearly we begin to discern the sinful distortions and hollow promises of free sin.

Nevertheless, flashy visuals are especially potent forces in our lives, Alastair Roberts explains, because the eye is especially susceptible to the "spectacular immediacy" of a stunning photograph online, for example. Ears are far less prone to this distraction, because the most powerful sonic realities are less immediately spectacular than visual realities, he argues.[18]

Again, the Christian priority on the invisible does not render the visible creation worthless.[19] It means that *what we see* is given its fullest meaning by *what we cannot see*. The physical gifts we enjoy seemingly are "thickened" by our capacity to see and treasure the unseen Giver.

All of this is mysterious to the world, but invisible realities govern our consuming. We are all hungry, thirsty, and needy for sustenance outside of us, but we give our attention and wealth to trying to satisfy our most essential longings with the goods and the vices so easily tapped on our phones. Therapeutic materialism is a scam. We order boxes of new goods that will never heal us and we buy bags of comfort food that will never truly comfort us, all because we are blind to the free gifts of God offered in his Son, Jesus Christ, whose body and blood have been given to us to sustain our eternal life and to feed

---

17. Observe this principle in reverse in Rom. 1:18–32.

18. Alastair Roberts, interview with the author via email (Jan. 23, 2016).

19. In one sense, the Christian priorities of faith, hope, and love bind together the visible and invisible. In love, we draw near to those we see, such as our neighbors. But in faith and hope, our love is properly grounded in the reality that our neighbors must finally be alienated from God or reconciled to God, and that forever. So our love (visible) takes on a particular hue because we see in them an eternity (invisible) that they perhaps cannot even imagine. The embodied/disembodied, visible/invisible, and tangible/intangible all work together in God's holistic ecosystem for the flourishing of his children. The Spirit, the water, and the blood all testify together (1 John 5:8).

the flourishing of our unceasing joy.[20] Jesus quenches the deep thirst that consumerism cannot slake.

Like a head-on collision of freight trains, the gospel of consumerism and the gospel of Christ smash:

- The gospel of consumerism says: everything you could possibly imagine for your earthly happiness and comfort is available in a dozen options, sizes, colors, and price points.
- The gospel of Jesus Christ says: everything you could possibly need for your supreme joy and eternal comfort is now invisible to the human eye.

In Christ, whenever we weigh the importance of anything in our lives, we weigh what is seen on one side, but it is outweighed by "an eternal weight of glory" on the other side.[21] In vivid theological parlance, the life of faith is about comprehending astonishing spiritual realities, which "requires a robust eschatological imagining, a faith-based seeing which perceives what is *not yet* complete—our salvation—as *already* finished, because of our union with Christ. It is a matter of seeing what is present-partial as future-perfect."[22] In common language, the life of faith is about comprehending the whole when we can only see a fraction. This is the work of imagination.

In an age of abundant visual vices and stunning, CGI-driven digital visual feats, the Christian imagination is starving for solid theological nourishment, warns theologian Kevin Vanhoozer. "Images are simply the icing on the cake of imagination, but there's little nutritional value in sugar," he told me. "The meat and potatoes of the imagination, the really nurturing part, involves words: in particular, stories and metaphors. To make sense of a metaphor, or to follow a story, is to make connections between things and, at the limit, to build a world."[23]

---

20. Isa. 55:1–2; John 6:25–59; 2 Pet. 1:3–4; Rev. 22:17.
21. 2 Cor. 4:17.
22. Kevin Vanhoozer, *Pictures at a Theological Exhibition: Scenes of the Church's Worship, Witness and Wisdom* (Downers Grove, IL: IVP Academic, 2016), 237, emphases original.
23. Kevin Vanhoozer, interview with the author via email (Feb. 26, 2016).

VIDEO LITERACY

Visual sugar cascades into our lives. By the end of 2015, close to five hundred hours of new video content was being uploaded to YouTube every minute of the day. Every minute! "As earlier ages moved from *orality* to *literacy*, we may be witnessing a tectonic cultural shift to *videocy*. We may not be programmers, but we make up what we could call the *digitality*: we are people of pixels."[24]

We also are witnessing a seismic social transition from passive video consumption to active (and hyperactive) video filming, editing, and sharing—all from our phones.[25] Video literacy is on the rise, and it comes with potent cultural force.[26] Many Christians are finding fresh ways of delivering edifying content in YouTube channels to serve the spiritual needs of their followers. We praise God for this. But as many Christians redeem video for God's glory, many others simply reflect the superficial appearances of the world.

It is worth reminding ourselves that the substance of our hope is not found in the latest visible spectacles on our glowing rectangles. Instead, our hearts delight in and relish a Christ we cannot yet see, a Christ we take by faith, a Christ who is so true and so real to us that we are filled in moments of this life with a periodic and expressive joy that is full of glory. Our imaginations must come alive to Christ so that we can "see" that we live in him, so that we can turn away from the visual vices grabbing our eyes, and so that we can live by faith and share a present joy as we anticipate the unimaginable future joy of his presence.[27]

HOPE AND BOREDOM

In the end, I wonder if most of the self-destructive patterns in our lives—from overeating to worrying to fighting to overspending to

---

24. Ibid.

25. Television and video have evolved quickly from live-linear viewing (traditional television) to on-demand streaming (archived movies, shows, sporting events, and YouTube videos) to semilive streaming (expiring videos such as recent shows and Snapchat videos) to real-live streaming (live sports and television and live personal video recorded on phones).

26. See Clive Thompson, *Smarter Than You Think: How Technology Is Changing Our Minds for the Better* (New York: Penguin, 2013), 83–113.

27. 1 Pet. 1:8–9; Jude 24–25.

grabbing our phones first thing in the morning—are the result of starved imaginations, malnourished of hope. When we live for what is visible and ignore what is invisible, we illustrate the definition of faithlessness. True faith lives for what is invisible and undisclosed. Every generation of the church faces its own unique struggle to focus on God and on the things not seen. The struggle is real—whether it is with the latest iPhone or the ancient household idol.

When I grow bored with Christ, I become bored with life—and when that happens, I often turn to my phone for a new consumable digital thrill. It is my default habit. "To become habituated to an iPhone is to implicitly treat the world as 'available' to *me* and at my disposal—to constitute the world as 'at-hand' for me, to be selected, scaled, scanned, tapped, and enjoyed."[28] In our phones, the digital age and the consumerist age merge, and our screens offer us everything we can see or desire, even "anonymous" compulsions and lurid fantasies.

DE-VICED

In light of the pace of all these digital temptations, a young man who struggled with digital vices asked if he should give up his smartphone and revert back to a "dumbphone." John Piper applied a wise strategy: "My guess is that some are going to say, 'Well, look, Piper, since the phone is not the problem, but the heart is the problem, it is pointless to pitch the phone.' To which I respond, 'No, it is not pointless to pitch the phone.'" We fight on two fronts in the battle for holiness in the digital age, he explained. "We are fighting on the *internal front* of the heart—the heart front to be so satisfied in Jesus, to see him so clearly and love him so dearly and follow him so nearly that nothing, not even a smartphone, can control us. But biblically, we are also fighting on the *external front* to remove or avoid stumbling blocks to our faith."

Then he concluded, "True freedom from the bondage of technology comes not mainly from throwing away the smartphone, but from

---

28. James K. A. Smith, *Imagining the Kingdom: How Worship Works* (Grand Rapids, MI: Baker Academic, 2013), 143, emphasis original.

filling the void with the glories of Jesus that you are trying to fill with the pleasures of the device."[29]

Our challenge in the digital age is twofold:

1. On the external front: Are we safeguarding ourselves and practicing smartphone self-denial?
2. On the internal front: Are we simultaneously seeking to satisfy our hearts with divine glory that is, for now, largely invisible?

Online allurements will always be with us, in a flood of cheap temptations, sexually charged images, and lurid ads. We must instead fill our hearts to the brim with glory so that our eyes learn to supernaturally scroll past the vapid images that naturally appeal to our eye lusts. To live an abundant life in this insatiable consumer society, we must plead in prayer for God-given power to turn our eyes away from the gigs of digital garbage endlessly offered in our phones and tune our ears to hear sublime echoes of an eternal enthrallment with the transcendent beauties we "see" in Scripture.[30]

---

29. John Piper, interview with the author via Skype, published as "When Should I Get Rid of My Smartphone?" Desiring God, desiringGod.org (Aug. 25, 2015), emphases added.

30. Ps. 119:18, 36–37.

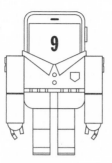

WE LOSE MEANING

The average output of email and social-media text is estimated at 3.6 trillion words, or about thirty-six million books—typed out every day! In comparison, the Library of Congress holds thirty-five million books.[1]

We now live in an information deluge only dystopian novelists could have foreseen. In the introduction to his landmark book, *Amusing Ourselves to Death*, Neil Postman contrasted two very different cultural warnings, those of George Orwell's *1984* and of Aldous Huxley's *Brave New World*. Orwell argued that books would disappear by censorship; Huxley thought books would be marginalized by data torrent. Postman summarizes the contrast well. "Orwell feared those who would deprive us of information. Huxley feared those who would give us so much information that we would be reduced to passivity and egoism. Orwell feared that the truth would be concealed from us. Huxley feared that the truth would be drowned in a sea of irrelevance."[2] Huxley seems to have won.

Reminiscent of Huxley and Postman, more recently, Pope Francis dropped his own warning about info overload in an encyclical on

---

1. Clive Thompson, *Smarter Than You Think: How Technology Is Changing Our Minds for the Better* (New York: Penguin, 2013), 47.

2. Neil Postman, *Amusing Ourselves to Death: Public Discourse in the Age of Show Business* (New York: Penguin, 1985), vii–viii.

global pollution, warning that "when media and the digital world become omnipresent, their influence can stop people from learning how to live wisely, to think deeply, and to love generously. In this context, the great sages of the past run the risk of going unheard amid the noise and distractions of an information overload." He argued that digital distractions must be held in check because true wisdom is the result of deep reading, self-examination, and "dialogue and generous encounter between persons." Merely amassing data, he warned, "leads to overload and confusion, a sort of mental pollution."[3]

JUNK FOOD FOR THE SOUL

Postman, Huxley, and the pope all share a techno-pessimism that I don't. And if info overload in the digital age is a problem, it strikes me as a secondary problem, one that I find somewhat limiting and unsatisfying as a full explanation, as if it doesn't reach the heart of the true problem.

First, declining literacy rates became a notable problem before Facebook was invented. As Oliver O'Donovan confessed to me: "My impression is that the damage to literacy is something of a *fait accompli* [irreversible reality], for which the electronic media are usually blamed. There are other factors at work, too. Literacy was not in wonderful health before the 1990s."[4] We must not assume that pre-smartphone generations were advanced in literacy and therefore more skilled in parsing out ultimate truth.

Second, the bigger challenge for us in the digital age is not the mental pollution of information overload, but the nutritional deficiency of the content that has been engineered, like modern snacks, to trigger our appetites. Online information is increasingly hyperpalatable, akin to alluring junk food. Breaking news, tabloid gossip, viral memes, and the latest controversies in sports, politics, and entertainment all draw us to our phones as if they were deep-fried

---

3. Pope Francis, "Encyclical Letter, Laudato Si' of the Holy Father Francis on Care for Our Common Home," The Holy See, w2.vatican.va (May 24, 2015).

4. Oliver O'Donovan, interview with the author via email (Feb. 10, 2016).

Twinkies held out on sticks at the state fair. Digital delicacies are eye-grabbing and appealing, but they lack nutrition.

Third, says Alastair Roberts, our phones make it possible to share and consume a steady diet of information that is pointless beyond making us feel connected to others. This is *phatic* communication—trivial knowledge that is shared to maintain some sort of social bond, but not to convey ideas (more on the pros and cons of digital "small talk" in chapter 12). Social media comes with an implicit contract, a sort of back-and-forth approval code that, over time, can erode the value of the information we share. I will follow you and "like" what you produce if you turn around and do the same for me. Inevitably the substance of our content can diminish, because the impetus for likes and shares is driven more by obligatory social reciprocity.[5]

So our problem is deeper than information overload; it is "our ungoverned appetite for connectedness with the immediacy and insistent urgency of the 'great communicative drama' of our society."[6] Our phones draw us into unhealthy habits not because we want unlimited information, but because we want to stay relevant and entertained. We want to be humored and liked. These social realities dwarf my concern over info overload.

"BREAKING NEWS"

Driving our desire to connect is our appetite for novelty. To make the point, imagine news agencies, once spread out in their own regions and distributing news to their audiences via television towers, radio transmitters, and bundles of magazines and newspapers. In the digital age, our news is increasingly confined to one big castle fortress (the web), with a few powerful gatekeepers that decide when to let news out and who sees it (increasingly, social-media platforms). Social media is not replacing the mass media; it is becoming the filter

---

5. Cal Newport, *Deep Work: Rules for Focused Success in a Distracted World* (New York: Grand Central, 2016), 208.
6. Alastair Roberts, interview with the author via email (Jan. 23, 2016).

through which the content produced by the mass media must now pass to reach untold masses.

If something is newsworthy, Twitter and Facebook will surely let us know. Between 2013 and 2015, Americans said social-media platforms were increasingly where they got their news; there were marked jumps among users of Twitter (52 percent to 63 percent) and Facebook (47 percent to 63 percent).[7]

Whether it's a "breaking-news" alert, a direct-message prompt, a text message, or a news app, our phones make our lives vulnerable to the immediacy of the moment in a way unknown to every earlier generation and culture. Social media and mobile web access on our phones all drive the immediacy of events around the world into our lives. As a result, we suffer from neomania, an addiction to anything new within the last five minutes.

Driven by social media, every media outlet races to the scene of the latest event. This feeds the "just-having-come-to-be" nature of news, writes O'Donovan. So-called urgent "breaking news," made hyperpalatable in social media, is the key to successful attention-grabbing by major platforms: "Devoting their full attention to the breaking wave, they echo its roar to us; we call upon them to show us the world new every morning, as though there never was a yesterday. Proverbially, news was thought to be refreshing," he says, echoing Proverbs.[8] Occasional good news can refresh us, but with our phones, even tragic news rushes at us in real time. And we welcome it. "What is striking about the speedy and wide-ranging communications of modern news is how on edge we are about them, as though we were constantly afraid that the world would mutate behind our backs if we were not *au courant* [current] with a thousand disassociated new pieces of information. This is a measure of our metaphysical insecurity, which is the engine of our modern urge for mastery."[9]

---

7. Michael Barthel et al., "The Evolving Role of News on Twitter and Facebook," Pew Research Center, journalism.org (July 14, 2015).

8. Oliver O'Donovan, *Ethics as Theology*, vol. 2, *Finding and Seeking* (Grand Rapids, MI: Eerdmans, 2014), 234. See Prov. 25:13, 25, along with 13:17.

9. Ibid., 235.

Hyperpalatable junk or not, we hate missing out (as we'll see in the next chapter). In our desire to "master" the world, we are made especially susceptible to novelty and prompts—we get texts, read tweets, or see notifications on our phones, and everything in our lives must stop. But in contrast to this immediacy, and the breaking news of the moment, "the steadfast love of the LORD never ceases; his mercies never come to an end; they are new every morning" (Lam. 3:22–23). The morning is when we "look back intelligently and look forward hopefully," writes O'Donovan. And yet, "the media's 'new every morning' (quickly becoming 'new every moment') is, one may dare to say, in flat contradiction to that daily offer of grace. It serves rather to fix our perception upon the momentary now, preventing retrospection, discouraging deliberation, holding us spellbound in a suppositious world of the present which, like hell itself, has lost its future and its past."[10]

Such an incredibly strong warning is appropriate if we, as eternal beings, live broken off from time by daily news cycles and disconnected from our place in God's story. We lose our place in history (as we will see later). And we lose our grip on ultimate meaning.

TREASURING WISDOM

Whether our greatest problem is the glut of information or the hyperpalatability of content, we must not shrug our shoulders (passivism), bend over our own reflections (narcissism), or fall into the pit of existential despair by disregarding our past and future (nihilism).

The solutions to all three core problems in the digital age are given by King Solomon—with prophetic warnings about an information age he never could have imagined. In his own day, as he looked at the proliferation of wisdom literature from all the world's sages, he saw benefit and value, but he also saw inundation. Sages will never stop writing books, he said, and we will never stop wanting to keep up. If we try to stay current, however, we will grow weary, because the accumulated

---

10. Ibid., 237.

libraries of wisdom have no end, and "much study is a weariness of the flesh."[11] Lacking self-control over the volume of our data ingestion introduces burdens that our physical bodies cannot carry.

That's where Solomon's three solutions come in.

First, in all the noise, Christians must identify and cherish wisdom. Before warning his son about the endless making of books and the weariness of much study, he wrote: "The words of the wise are like goads, and like nails firmly fixed are the collected sayings; they are given by one Shepherd. My son, beware of anything beyond these" (Eccles. 12:11–12a). We must assign a value judgment to all information we take in. We don't engage with digital content simply to keep up, to be informed, or to connect. Instead, we plug our ears to the noise of novelty so that we can identify meaning and embrace truth, goodness, and beauty. We now live in the golden age of quality and edifying online content, made available free of charge. But do we slow down and absorb these sites with the value they represent, or do we lose the value of these sites in the clamor of immediacy, the rapidity of some invisible expiration date, and the hyperpalatability of all the other digital noise?

Cherishing wisdom is a discipline of literacy. "What literacy used to mean was a capacity to interrogate an appearance, including the appearance of numbers. What do they mean? What is the lived experience behind them?" Literacy asks: What's the point? "Perhaps the greatest threat we face is that of living with short attention-spans, caught now by one little explosion of surprise, now by another. Knowledge is never actually given to us in that form. It has to be searched for and pursued, as the marvelous poems on Wisdom at the beginning of Proverbs tell us."[12] Without wisdom, we foolishly get lost in the aimless *now*, in the explosion of novelty. Without wisdom, we foolishly get unhitched from our *past* and from our *future*.

Second, in all the noise, Christians must strive for fearful obedience over frivolous information. After his statement about the end-

---

11. Eccles. 12:12.
12. O'Donovan, interview with the author via email (Feb. 10, 2016).

lessness of books, Solomon wrote: "The end of the matter; all has been heard. Fear God and keep his commandments, for this is the whole duty of man" (Eccles. 12:13). More important than information access, more valuable than social-media prominence, is Godward obedience.

Third, in all the noise, we must embrace our freedom in Christ, as we step back from the onslaught of online publishing and the proliferation of digital sages. By grace, we are free to close our news sources, close our life-hacking apps, and power down our phones in order to simply feast in the presence of friends and enjoy our spouses and families in the mystery, majesty, and "thickness" of human existence.[13]

TECHNOLOGY AND WISDOM

Going back to the definition I used at the beginning of the book, *technology* embraces more than our smartphones. Adam and Eve were created, naked, to live in an earth full of animals. As an initial nudge toward technological progress, God invented the first textiles and the first sword.[14] Beginning with that first attire and first blade, everything else that would be woven, mined, smelted, machined, polished, and mass-produced fits under the *technology* umbrella.

Job 28 is a poem celebrating the technological innovation of man. We can scour the planet for raw materials such as iron and copper. We can go where birds, animals, and even travelers have never been. We can venture down into dark, echoing shafts under the earth and swing back and forth on the ends of ropes as we are lowered deeper and deeper to extract gold flakes and diamonds. We can overturn mountains by the roots.

---

13. Eccles. 9:7–9. Dietrich Bonhoeffer: "To be sure, an excessive cultivation of human relationships . . . lead[s] to a cult of the human that is disproportionate to reality. In contrast to that,
what I mean here is simply that people are more important to us in life than everything else. That
certainly does not mean that the world of material things and practical achievements is of less
value. But what is the most beautiful book or picture or house or estate compared to my wife,
my parents, my friend? Yet the only person who can speak this way is one who has really found
human companionship in life. For many today, people are nothing more than part of the world of
things." Dietrich Bonhoeffer, *Letters and Papers from Prison*, ed. Christian Gremmels, trans. Isabel
Best, vol. 8, *Dietrich Bonhoeffer Works* (Minneapolis: Fortress, 2010), 509.

14. Gen. 3:21, 24.

If Job 28 is a glorious hymn to celebrate the innovation of man (vv. 1–11), it is also a warning song about the limits of the wisdom we can find by our devices (vv. 12–28). When it comes to searching out the meaning of our existence in this world, all of our technology cannot take us deep enough or high enough. True wisdom is beyond the reach of our pickaxes and techniques. We can climb down into the dusty, dark shafts that go deep into the earth, but wisdom is not there. We can go under the sea, but wisdom is not there, either. All the rich gold brought into the light will not disclose wisdom. We can be tech-savvy fools.

In this digital age of overwhelming content, we must not relinquish ourselves to passivity or to egoism. And we certainly must not drown in a sea of irrelevant news and gossip. Instead, we must learn to treasure what is most valuable in the universe—God. When we turn to God, we find that the most precious wisdom and knowledge is not hidden under a mountain or embedded in the newest device, but found in Jesus Christ.[15] He defines the purpose and meaning of all life. He orients what is truly important and valuable for us in the digital age, and in every age.

---

15. Col. 2:3.

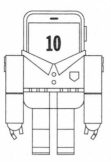

WE FEAR MISSING OUT

Missing a potential spouse, missing a perfect job offer, missing a golden stock tip, or missing a party with our friends—missing out leaves a sting of regret we all hate. Foresight is blurry, but hindsight is 20/20, and that means we remember our past misses with crystal clarity. When we miss out too many times, we can begin to dread the next miss inordinately.

So our phones and social media serve as a real-time refresh of our comparisons with the lives of others, constantly feeding our "fear of missing out" (FOMO). FOMO and social media go hand in hand. Even the new entry in the *Oxford English Dictionary* confirms the link: "FOMO—fear of missing out, anxiety that an exciting or interesting event may be happening elsewhere, often aroused by posts seen on a social media website."[1]

FOMO can be diagnosed through more basic symptoms of "disconnection anxiety," also known as "no-mobile-phone phobia"—nomophobia—the fret when we find ourselves prevented from accessing our digital worlds. This strain of FOMO is highly contagious and progresses rapidly. For example, one young woman, who was raised offline in an Amish community for eighteen years

---

1. "FOMO," *Oxford English Dictionary*, oed.com (June 2015).

before entering the online world, quickly caught the disconnection fever. After adjusting to the non-Amish life and adapting to digital America, she took an offline mission trip. "I was thinking, I just can't wait to go back to the U.S. where I can be connected to technology again and see what all is happening. Because it feels like I'm naked or something without being constantly updated on what's going on."[2]

If a former Amish woman who didn't touch an iPhone until adulthood, and who maintains relatively healthy online habits, is susceptible to this fear, I suspect many of my worst phone habits are born from FOMO. I want to know, I want to see, and I don't want to be left out.

My desire to never be socially left out comes at the price of beeps, pings, and endless feed refreshes. I constantly check my phone to make sure I'm not missing anything. But others also pay a price for my so-called "relevance." When it comes to cultural FOMO, we are eager to turn the tables and heap shame on others for not having yet ingested the movies, television series, or viral stories that we have already consumed. Whenever someone admits that they are behind on these cultural products, we are quick to expose them. "Much as we begrudge being on the receiving end of the guilt, we dish it out in equal portions."[3] Yes, we have blood on our hands—because we both carry and spread this vicious FOMO disease. It feels so good to flaunt our relevance over one another's irrelevance.

The sick irony is that our FOMO causes us to run right into the "just-having-come-to-be" nature of news, which only deepens the problems we addressed in the last chapter and stokes our fears about this world. "I think more than ever before Christians are news junkies," counselor Paul Tripp told me. "More than ever before, through social media and websites and a 24-hour news cycle, we are aware of what is happening around us. And I think for many of us this has

2. Olga Khazan, "Escaping the Amish for a Connected World," *The Atlantic* magazine (Feb. 17, 2016).

3. Kate Hakala, "There's a Special Kind of 'FOMO' Stressing Us Out—And We're Doing It to Ourselves," *Mic*, mic.com (May 21, 2015).

raised our fears."[4] Yes, to have smartphones is amazing, but to have the Internet on our phones is to also have immediate access to all of the world's major tragedies, sorrows, bombings, and acts of terrorism. Are we prepared to carry this burden?

CORE FOMOS

We can boil down our core online fears to two anxieties, says theologian Kevin Vanhoozer: "status anxiety (what will people think of me?), and disconnection anxiety ('I connect, therefore I am')." But connected to what, and at what cost? "I'm afraid that, for many, the answer too often is: connected to *the empire of the entertainment-industry complex*. We live in what has been described as an 'attention economy,' and the Sunday morning sermon seems weak in comparison to an Internet-surfing session. The latter allows us to ride the waves of popular culture and opinion." Like so many points throughout this book, it's a question of meaning. "The sobering question for the disciple is whether our attention is being drawn to something worthwhile. Spectacles are ephemeral, which is why those who suffer from FOMO are always on the lookout for The Next Big Thing. Disciples who are awake to reality have their attention fixed on the only breaking news that ultimately matters, namely, the news that the kingdom of God has broken into our world in Jesus Christ. This breaking news demands our sustained attention, and a wide-awake imagination."[5]

Christians, perhaps like never before, are tempted to remain tethered to the daily news cycle, viral videos, political forecasts, and entertainment gossip (as we saw in the last chapter). Our hyperconnection is fueled by our FOMO. We hate being left out, so we focus on every Next Big Thing, such as the upcoming blockbuster film. And we forget about big, glorious realities like the inbreaking new creation of God.

4. Paul Tripp, interview with the author via phone, published as Paul Tripp, "God's Glory Must Enchant Us," Desiring God, desiringGod.org (Feb. 1, 2016).

5. Kevin Vanhoozer, interview with the author via email (Feb. 26, 2016).

SADNESS AND SILENCE

Smartphone FOMO is a universal experience, and as you can see, Christians are not immune. When writer Andy Crouch took forty days offline—no screens and no social media—he said the experience was mostly delightful. "But I will say this: FOMO—the 'fear of missing out'—is a real thing," he admitted. "What I was most afraid of missing out on was not information, but affirmation. I discovered how attached, or maybe addicted, I was to the small daily dose of reassurance that other people 'like' me and 'follow' me. . . . It was sobering how strong the pull was on me."[6]

This desire for personal affirmation is perhaps the smartphone's strongest lure, and it is only amplified when we feel the sting of loneliness or suffering in our lives. At the first hint of discomfort, we instinctively grab our phones to medicate the pain with affirmation. This habit could not be more damaging.

What we often forget when we scroll social media is how "professionalization" shapes our public platforms. Most of us know that our present and future employers will likely review what we've published on Twitter, Facebook, and Instagram. This employer omniscience is daunting and sometimes intimidating, but it means that whatever we put in our public feeds tends to look edited, polished, and "with it"—in control, confident, and sure. Our social personas are increasingly conditioned by corporate expectations.[7] But when suffering hits, we forget that social media calls for a one-dimensional, carefully manicured projection of the self. Then we trudge our sorry selves to social media in order to confirm just how awful our lives are compared with everyone else's togetherness!

In other words, FOMO plays insidious mind tricks when our sorrows are prolonged. When a sense of pain or suffering hits, we turn to

6. Cited in Joshua Rogers, "Five Questions With Author Andy Crouch," *Boundless*, boundless.org (June 15, 2015).

7. See Donna Freitas, *The Happiness Effect: How Social Media Is Driving a Generation to Appear Perfect at Any Cost* (New York: Oxford University Press, 2017), and Ariane Ollier-Malaterre, Nancy P. Rothbard, and Justin M. Berg, "When Worlds Collide in Cyberspace: How Boundary Work in Online Social Networks Impacts Professional Relationships," *Academy of Management Review* (Jan. 2, 2013).

our phones—and by turning to our phones, we exacerbate the pain, explains pastor Matt Chandler, a survivor of brain cancer. Imagine someone enduring prolonged suffering or depression, sitting at home in his or her pajamas. "You crawl into bed, and you grab your phone. You start scrolling through your Instagram account. Here's what you find: everybody's marriage is awesome. Their kids are incredible. They're counting money. And they don't struggle at all. There's no pain. There's no sorrow. And here you are in your trial. You ate a whole gallon of ice cream watching a series on Netflix. You start to resent them. You start to grow in anger against them. 'Really? Me, Lord? I'm enduring this trial? What about them?' In your trial, your insidious, wicked heart will be exposed, and comparison is how it plays itself out."[8]

FOMO ENVY

Comparing ourselves is a social evil that thrives among socioeconomic peers. Among such peers, envy is not merely wanting what others have, but wanting it *because* they have it. Or it can manifest in the desire that they *not have* what I cannot have. This sin insidiously aims at destroying others' goods and gifts in light of my own loss and lack.

In other words, envy thrives by concrete markers of comparison, writes Brad Littlejohn. We carry envy into the world of social media on our phones, where we "can readily tabulate how many 'likes,' how many comments, how many 'favorites,' how many 'retweets' or 'repins' our friend's status/picture/tweet/post received, versus how many ours received. To the envious heart, each one of these little icons of approval is a red-hot poker, stoking the burning fire of bitterness and envy," he says. "The envious heart will masochistically store up each painful reminder of the others' success, tabulating them and rehearsing them, until it seems like the whole world is conspiring against it."[9] That may

---

8. Matt Chandler, sermon, "James: Trials/Temptations," The Village Church, thevillagechurch .net (Feb. 15, 2015).

9. Brad Littlejohn, "The Seven Deadly Sins in a Digital Age: V. Envy," *Reformation 21*, reforma tion21.org (Dec. 2014).

be an extreme example, but even lesser forms of envy crush our joy and squeeze the life out of our souls under the weight of comparison. The accumulated shares, likes, and friends offer us an irresistible place for comparison. Perhaps it is not far off to say that "Facebook is the CNN of envy, a kind of 24/7 news cycle of who's cool, who's not, who's up, and who's down."[10] In effect, social media becomes a bellows that keeps pumping fuel into the internal fire of our envy.

All of this FOMO-driven envy, sparked by personal suffering and stoked with the lure of personal affirmation, is a combustible pile of chaff.

THE BIRTHPLACE OF FOMO

FOMO is neither unique nor modern. It predates the acronym coined in 2004, it predates WiFi, and it predates our smartphones. FOMO is an ancient phobia with a history that reaches back far before we started using our opposable thumbs to text one another gossip. We can say that FOMO is the primeval human fear, the first fear stoked in our hearts when a slithering Serpent spoke softly of a one-time opportunity that proved too good to miss. "Eat from the one forbidden tree, Eve, 'and you will be like God.'"[11]

What more could Eve or Adam want—to escape creaturehood, to become their own bosses, to preserve their own independence, to define their own truth, to become all-knowing, and to delight in autonomous regality. They could keep all the glory for themselves by becoming gods and goddesses! Who could refuse the irresistible chance to become godlike in one bite?

These words—this lie!—were loaded with a succulent promise too good to be true. It was false flattery. It was Satan's attempt to dethrone God by spinning words into an insurrection by God's own image bearers. In other words, FOMO was Satan's first tactic to sabotage our relationship with God, and it worked. And it still does.

Behind the first sin was a desire for a "different" life. We can all

---

10. Freitas, *The Happiness Effect*, 39.
11. See Gen. 3:5.

imagine better lives, yes, and in the words of one novelist, "sometimes I can hear my bones straining under the weight of all of the lives I'm not living."[12] The strain of living just one life is enough, but give yourself time to think about all the other lives you could be living, and the weight of possibilities will press down and lure you to a mirage of escapism just as it did for Adam and Eve. This is FOMO.

But FOMO did not end under a forbidden tree in Eden. It only started there, kindling into a forest fire of FOMO that has never since been extinguished in the human experience. Every day, sinners are still animated by the empty promise of reaching some level of self-sufficiency where God will be finally rendered unnecessary.

Every day, we are faced with the lives that we cannot live, the lives that only others can live, and the lives that God has explicitly forbidden us to live. By insisting that we, God's creatures, are missing out, the lies of FOMO make us easy targets for advertisers; sharpen the sting of our quarterlife and midlife crises; and sour the elderly years, when the reality of cultural "missing out" becomes most obvious.

FOMO IN THE GRAVE

One of the longest-running FOMO object lessons is told by our Savior in Luke 16:19–31, a story of contrasts between eternal loss and eternal glory. The story begins with a rich man (who seems to not be missing out in any social or financial way) and Lazarus, a poor man (who seems to be missing out in every way imaginable). Their contrasts are merely temporary, because both men die and face eternity.

The story of the rich man and Lazarus is the grand story of role reversal. By the end, we find a former rich man (who has lost everything) and a former beggar (who has gained everything). The former rich man now faces *eternal torment* as a beggar who pleads for a drop of water to cool the agony of judgment. The beggar now finds *eternal delight* as a redeemed sinner whose regrets and fears have been washed away in the eternal joy of God's restorative presence.

---

12. Jonathan Safran Foer, *Extremely Loud and Incredibly Close* (Boston: Mariner Books, 2005), 113.

At this point, the rich man (now the eternal beggar) is missing out (MO), and he fears that his loved ones will, too (FOMO). His urgent plea to Abraham: resurrect the beggar Lazarus and send him back into the world to tell the rich man's five brothers of eternal life, so that they will hear and believe, and thereby escape this wretched eternal missing out. This is the rich man's desperate cry.

Jesus makes the moral of the story obvious. Where God's Word is opened, read, and embraced by the hearer, there is no eternal fear—only the promise of eternal restoration for everything missed out on in this life.

ONE LEGIT FOMO

As this story highlights, one legitimate FOMO cuts through all the other FOMOs of life: the fear of *eternally* missing out. God's wrath is real. And apart from Christ, there is only eternal destruction. The wealthy man in Jesus's parable is a portrait of life's greatest tragedy— a man filling his pockets, his belly, and his life with vain pleasures. He bought Satan's old lie to Eve, choosing the foolish path of God-ignoring self-sufficiency, and never embraced God as his greatest treasure. He deadened the reality of judgment with the Novocain of self-indulgence, and by it he destroyed himself eternally.

In this condition of unbelief, the rich man faced the agony of the one most dreaded missing out, an eternal missing out, a weeping-and-gnashing-of-teeth missing out. "Therefore, while the promise of entering his rest still stands, let us fear lest any of you should seem to have failed to reach it" (Heb. 4:1). The fear of missing out on eternal life is the one FOMO worth losing sleep over—for ourselves, our friends, our family members, and our neighbors.

But if you are in Christ, the sting of missing out is eternally re-moved. FOMO-plagued sinners embrace the gospel of Jesus Christ, and he promises us no eternal loss. All that we lose will be found in him. All that we miss will be summed up in him. Eternity will make up for every other pinch and loss that we suffer in this momentary life. The doctrine of heaven proves it. The new creation is the restoration of everything broken by sin in this life; the reparation of everything we lose in this world; the reimbursement of everything we miss out on in our social-media feeds.

Lazarus learned this blessed truth: heaven is God's eternal response to all of the FOMOs of this life. Heaven will restore every "missing out" thousands of times over throughout all of eternity.[13] Therefore, the motto over the allurement of the digital age is set in the slightly altered words of the apostle Paul: I count every real deprivation in my life—and every feared deprivation in my imagination—as no expense in light of never missing out on the surpassing worth of knowing Christ Jesus my Lord for all eternity.[14]

---

13. Acts 3:21.
14. See Phil. 3:8.

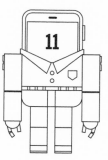

WE BECOME HARSH TO ONE ANOTHER

What should I do with the dirt I have on you? That's a question we all face at some point.

While there are many "one anothers" in the Bible, "compare one another" is not one of them, and yet this is the direction we tilt online. We celebrate celebrities. We disdain nobodies. With those most like us, we grow envious and harsh. We live between façades of online confidence that resemble flimsy stage sets. "Social media—as the current system of numbers and money dictates—is not genuine life," writes Essena O'Neill, the former Instagram model we met earlier. "It's purely contrived images and edited clips ranked against each other. It's a system based on social approval, likes and dislikes, validation in views, success in followers. It's perfectly orchestrated judgment."[1]

We go online to compare one another. We chide one another. We become jealous of one another. And when we get dirt on one another, we fall into perfectly orchestrated judgment against one another.

And there's always an app for that.

1. Essena O'Neill, "Social Media Addiction and Celebrity Culture," letsbegamechangers .com (Oct. 30, 2015). This quotation from Essena O'Neill appeared in materials on her website, letsbegamechangers.com, at the time of writing. Prior to publication, that site was taken down. Interested readers can find the quotation by searching letsbegamechangers.com through web .archive.org.

PEEPLE

The creepily named app Peeple originally was designed to offer users the chance to calculate one- to five-star ratings of the people they know—friends, coworkers, and former romantic partners. We are not talking about critical reviews of bad restaurants or defective products; we are talking about public evaluations of private individuals.

What could go wrong with that? Well, a lot. The *Washington Post* called Peeple "inherently invasive," "objectifying," "reductive," and a source of stress and anxiety for "even a slightly self-conscious person." Even more, Peeple generated a platform that would encourage invasion of privacy and even harassment. At the very least, it produced the feeling of "being watched and judged, at all times, by an objectifying gaze to which you did not consent."[2]

So Peeple's developers returned to the drawing board and rethought their policies and procedures to ensure that the site functioned more to promote good people than to denigrate villains. Open-platform evaluations always tend toward the destructive, as we instinctively know.

Besides apps like Peeple, our phones provide many windows into this harsh reality. We see condescending comments on articles. We see snarky, judgmental remarks on Facebook. We see jolting tugs-of-war on Twitter. We see accusations about evangelical leaders on blog posts. No matter where the skirmishes start, they evidence an often endless (and loveless) war. Whether we find ourselves on the sidelines or front lines of these debates, we face a vital question: How should we handle the sins and weaknesses of people around us?

Thankfully, our script is written in Matthew 18:15–20, and it's clear: if a brother or sister in Christ sins against you in a serious way, go and tell him his fault in private. If he repents, an incredible restoration has unfolded in God's eyes, and the reconciliation is done. If he fails to repent, however, you next bring one or two witnesses along

---

2. Caitlin Dewey, "Everyone You Know Will Be Able to Rate You on the Terrifying 'Yelp for People'—Whether You Want Them to or Not," *The Washington Post* (Sept. 30, 2015).

to confront the wrongdoer. If that does not work, you should share the wrong with church leaders and then ultimately with the entire local church. If the wrongdoer refuses to repent, he is no longer to be treated as a brother in Christ.

There's a process for this discipline, and it is based on brotherly love, not guerrilla warfare. Likewise, there's a process for confronting church leaders who have sinned, and it begins with a method to authenticate accusations and then calls for sins to be addressed in accordance with denominational processes and trials.[3] In every case, Scripture—not social tools—guides the process.

CALLING

When it comes to confronting the sin of any believer or pastor in our lives, the private, scriptural process must be respected, even when it unfolds slowly. The key to the entire process is *calling*—a few people who are in proximity are *called in* to address a certain case.[4] Sins and failures should be handled face to face between the wrongdoer and the person wronged, along with the witnesses, all under the discretion of a local church.

For those of us who are not "called" into a situation (the majority of us), our script calls for us to take the very countercultural posture of self-restraint, of not talking about the sins in question.[5] We cover over sins, not so they can fester in silence, but so that those called to the situation can deal with those sins in the light of God's script. In fact, as the script makes clear, the conclusions of two or three believers who are called into a particular situation bear far greater weight in God's eyes than those of two or three hundred people filled with anger, frothing up one another in Facebook comments.

Our priority to honor God's design here stops us from texting friends to share the dirt we have on others. Such self-control is not

---

3. 1 Tim. 5:19–21.
4. *Calling* is an important point reinforced by the wise counsel of Puritan Richard Baxter, and much of his writing informs this chapter, especially *The Practical Works of the Rev. Richard Baxter* (London: James Duncan, 1830), 6:386–413.
5. Prov. 10:12; 11:12–13; 17:9.

intuitive, but it is imperative—and it is how we protect the honor of our neighbors and our brothers and sisters in Christ.

WHISTLE-BLOWING GONE WRONG

In a smartphone society, social media will continue to serve as a powerful tool for exposing fraud, toppling dictators, blowing the whistle on crimes, and recording and exposing racial injustices. For Christians, these tools will offer us means of advocacy and social justice,[6] and, when necessary, will serve in moments when it is essential to expose ongoing sin and false doctrine that would otherwise fester in silence in churches and denominations. But what at first appears to be a noble attempt to expose past sin often goes too far and leads to a collective online vendetta, even by Christians.

There's a very real temptation for those who are not called into a certain situation to attempt to judge cases remotely, make premature conclusions, and then attract an online groundswell of support. But crowdsourcing verdicts and spreading unfounded conclusions online can destroy the reputation of a Christian. This is when the script goes satanically wrong.

In an age when anyone with a smartphone can publish dirt on anyone else, we must know that spreading antagonistic messages online, with the intent of provoking hostility without any desire for resolution, is what the world calls "trolling" and what the New Testament calls "slander."[7] The verb form of the Greek word used in the New Testament literally means "to speak against." Online slander includes spreading false information and rumors about others. But biblical slander is slanderous for its end result: injured reputations.

---

6. See Heidi A. Campbell and Stephen Garner, *Networked Theology: Negotiating Faith in Digital Culture* (Grand Rapids, MI: Baker Academic, 2016). Hashtag advocacy can be a powerful tool, but it's not without limitations. See Malcolm Gladwell, "Small Change: Why the Revolution Will Not Be Tweeted," *The New Yorker* (Oct. 4, 2010). More important is giving blood, volunteering, helping a neighbor, and visiting orphans and widows in their affliction (James 1:27). Whatever the value of social-media advocacy, we must hold ourselves to the higher advocacy standards of Christ (Matt. 25:31–46).

7. See 2 Cor. 12:20; 1 Pet. 2:1 (καταλαλιά, "slander" or "slanders"); Rom. 1:30 (κατάλαλος, "slanderers"); James 4:11; 1 Pet. 2:12; 3:16 (καταλαλέω, "to slander" or "to speak against").

JAMES 4

In a chapter loaded with wisdom on how Christians are to handle the dirt they have on one another, we find slander: a sin that "violates the early Christian commandment because of its uncharitableness, rather than its falsity."[8] That's the key. Tim Keller and David Powlison define slander as "not necessarily a false report, just an 'against-report.' The intent is to belittle another. To pour out contempt. To mock. To hurt. To harm. To destroy. To rejoice in purported evil."[9]

Slander is not a public debate over ideas or a public rebuke of false teaching (more on that later). We can certainly debate ideas and doctrine in public as long as we are fair and principled, and represent our opponents' views with clarity and charity.[10] What James 4:11–12 warns against is "attacking a person's motives and character, so that the listeners' respect and love for the person is undermined."[11]

In his comments on James 4:11–12, written long before the advent of the iPhone, pastor R. Kent Hughes said: "Personally, I can think of few commands that go against commonly accepted conventions [slander] more than this. Most people think it is okay to convey negative information if it is true. We understand that lying is immoral. But is passing along damaging truth immoral? It seems almost a moral responsibility!" This is why the biblical definition of slander is countercultural to the smartphone generation. "By such reasoning, criticism behind another's back is thought to be all right, as long as it is true. Likewise, denigrating gossip (of course it is never called gossip!) is okay if the information is true. Thus, many believers use truth as a license to righteously diminish others' reputations."[12] What is done in the name of "exposing truth,"

8. On "λαλέω," see Gerhard Kittel, Geoffrey W. Bromiley, and Gerhard Friedrich, eds., *Theological Dictionary of the New Testament* (Grand Rapids, MI: Eerdmans, 1964), 4:4.

9. Tim Keller and David Powlison, "Should You Pass on Bad Reports?" The Gospel Coalition, blogs.thegospelcoalition.org (August 4, 2008).

10. For a further discussion on the characteristics of humble debate, see John Newton's principles in Tony Reinke, *Newton on the Christian Life: To Live Is Christ* (Wheaton, IL: Crossway, 2015), 256–59.

11. Keller and Powlison, "Should You Pass on Bad Reports?"

12. R. Kent Hughes, *James: Faith That Works*, Preaching the Word (Wheaton, IL: Crossway, 1991), 194.

with the single goal of undermining someone's character, is an expression of slander.

Unless confronted, James 4:11–12 warns, faultfinders and fault-spreaders eventually take their seats as rogue judges who stand *over* the law. In their impatience and cynicism regarding standards and processes, these faultfinders can *become* the law, judge, and jury in order to pronounce guilt and dispense retribution against a wrongdoer. Such impulses attract online mobs that can quickly heap collective shame. The act of exposing dirt on someone rarely stops with whistle-blowing and exposé, but typically moves quite naturally into a collective vendetta that leverages mass online outrage to see documentable harm done to the wrongdoer.

But God prevents the wounded from becoming the wounders. To do this, his script often cuts against the grain of conventional wisdom, and it always cuts against the impulses of our flesh. Humility calls us to follow a script of counterrevolution in the midst of a Wikileaks generation. In this age of Peeple ratings, whistle-blowing, and cover-up exposing, we have been given a countercultural script we must follow when dealing with the dirt we have on others.

COMMAND NINE

James 4 is really just a restatement of the ninth commandment,[13] a necessary command against lying about our neighbor inside a courtroom and a bold command that calls us, outside the courtroom, to be "more disposed to covering our neighbor's blemishes than to publicizing them."[14] As the Westminster Larger Catechism explains it, this is a call for "a charitable esteem of our neighbors; loving, desiring, and rejoicing in their good name; sorrowing for, and covering of their infirmities; freely acknowledging of their gifts and graces, defending of their innocence; a ready receiving of a good report, and unwillingness to admit of an evil report, concerning them; discour-

---

13. Ex. 20:16; Deut. 5:20. For a full treatment, see John M. Frame, *The Doctrine of the Christian Life (A Theology of Lordship)* (Phillipsburg, NJ: P&R, 2008), 830–43.

14. Michael Horton, *Calvin on the Christian Life: Glorifying and Enjoying God Forever* (Wheaton, IL: Crossway, 2014), 178.

aging talebearers, flatterers, and slanderers."[15] Again, it restrains us from spouting guesses about the motives and intentions of others.[16] Extreme caution and self-restraint are called for with the dirt of our online neighbors.

God wants us to practice the discipline of covering the sins of others in love[17] as we give them space for discipline (when needed) and for personal repentance.[18] We acknowledge the often unseen and invisible work of the Holy Spirit in the world to bring conviction of sin. And so we walk by faith, knowing that God is at work in his children.

To this end, I find it helpful to frequently recall the frank admission of Charles Spurgeon: "The easiest work in the world is to find fault."[19] Yes, and the tools to spread our findings have never been simpler or more powerful. A "quarrelsome man" who desires to ignite strife and fan it into a flame of contention will surely find his way to the kindling of social media. "With social media, we can now harm and embarrass and stigmatize people with greater force than ever before in human history," warns pastor Ray Ortlund. "Self-restraint has never been more important."[20] Each of us has an inner troll, an inner slanderer—some part of us that would love to text some dirt to a friend, publish dirt online, and anonymously consume that dirt online. "If 'the words of a whisperer are like delicious morsels' then online comments are like an all-you-can-eat buffet."[21] And who can fast in the presence of a buffet?

Our gluttonous fascination with the failures of others long predates social media. Faultfinding is an ancient hobby, meant to prop up a façade of self-importance, even among Christians. Faultfinding

---

15. Question 144.

16. See Thomas Boston, *The Whole Works of Thomas Boston*, vol. 2, *An Illustration of the Doctrines of the Christian Religion, Part 2* (Aberdeen: George and Robert King, 1848), 323.

17. Prov. 10:12; 11:12–13; 17:9; 1 Pet. 4:8.

18. Prov. 28:13; 1 John 1:8–10.

19. C. H. Spurgeon, *The Metropolitan Tabernacle Pulpit Sermons* (London: Passmore & Alabaster, 1910), 56:408.

20. Ray Ortlund, interview with the author via email (March 1, 2012).

21. Sammy Rhodes, *This Is Awkward: How Life's Uncomfortable Moments Open the Door to Intimacy and Connection* (Nashville: Thomas Nelson, 2016), 196. See Prov. 18:8.

destroys our love for others. Faultfinding runs contrary to Calvary. In Christ, our pardoned sins are plunged into a grave—but the slanderer keeps going at night to exhume his neighbor's sins in order to drag those decomposing offenses back into the light of the city square.[22] This is why, when Puritan Richard Baxter believed slander had reached epidemic proportions in the church of his own day, he confronted the sin—and paid the price. "My conscience, having brought me to a custom of rebuking backbiters [slanderers], I am ordinarily censured for it, as one that defends sin and wickedness."[23] Ouch. Censure faultfinders at your own risk.

We must have courage to turn away from online slander or to confront it as slander. We must have eyes to see through the hollow accusation that our silence is a passivity that allows sin to run unchecked. God knows that we will have dirt on our neighbors and on other Christians, and that's why he tells us what to do in his script. His Word tells us that it is *wrong* to slander, it is *wrong* to feed on slander, and it is *right* to confront the prevalence of the sin online (even if we incite slander for doing so!).

SHOULD WE CONFRONT A CHRISTIAN'S SIN ONLINE?

When handling serious personal sin and false teaching, we see two distinct scenarios in Scripture: *sins* inside a local church and *heresies* outside a local church. I'll set them together:

	Sins inside a Local Church*	Heresies outside a Local Church†	
○	These sins include major doctrinal errors, major moral failures, and persistent and schismatic divisiveness found within a local church, in its people or leaders.‡	These sins include false teaching published in persuasive books, articles from prominent writers, public sermons, seminary teaching, and denominational stances—errors with reach beyond one local church.	○
○	Call for private rebuke.	Call for public rebuke.	○

---

22. Baxter, *The Practical Works of the Rev. Richard Baxter*, 6:408.
23. Ibid., 6:393.

	Sins inside a Local Church*	Heresies outside a Local Church†	
o	Response begins with the person wronged, then other church members, church leaders, and finally the entire local church (if needed).	Skillful response to false teaching is an essential function of qualified church leaders, but here it seems best done by leaders with widely appreciated authority.	o
	Resolution ends in acquittal, repentance, or churchwide rebuke and possibly excommunication.	Resolution appears to end in public exposure.	

*. Matt. 18:15–20; 1 Tim. 5:19–20; Titus 1:9.
†. Gal. 2:7–14.
‡. D. A. Carson, "Editorial on Abusing Matthew 18," *Themelios*, themelios.thegospel coalition.org (May 2011). These are his three categories that qualify for Matthew 18 confrontation: major doctrinal error (1 Tim. 1:20); major moral failure (1 Corinthians 5); and "persistent and schismatic divisiveness" (Titus 3:10–11).

To me, these are the two clearest biblical categories. But in the age of social media, where digital voices can be collectivized, a third category emerges—one used to expose allegations of church cover-ups and to rebuke prominent Christian leaders for asserted moral failures. This third category calls for public accountability structures that supersede the authority of a local church or a denomination.

Public scandal may call for public rebuke, and the Bible does not hide from us the discomforting fact that scandalous pastors have existed and will exist in the future. And when scandal hits, a governing authority must actively step in with a timely, fair, and impartial investigation in order to acquit or punish, no matter the fallout.[24] When the scandal includes criminal accusations, civil authorities must be called in to protect and investigate. The church needs to carry out the first process in the fear of God and in private (in respect to James 4). It must allow the second process to proceed with cooperation and without hindering the truth (in respect to command nine). Yet in a fallen world, both of these processes are flawed, and sometimes, when ecclesiastical and civil authorities are aware and engaged, division among Christians cannot be prevented. Lingering

24. 1 Tim. 5:17–21.

questions and unresolved frustrations may hang over such a situation for many years, leaving hurts and tensions that call for the strongest form of faith—deep trust in God's sovereign will, his perfect timing, and his future verdict.

SHOULD I TWEET IT?

So what is my role in social media when church scandals arise? Understanding the wide complexity of situations,[25] and knowing my own propensity to backbite, before I grab my phone, I must cautiously ask myself:

- Would my action violate James 4 or command nine?
- Would my action obstruct or render God's accountability structures in a local church useless or disrespect the slow and careful pace of a denomination already alerted to the situation?
- Given my proximity to or distance from the situation, has God called me to write, comment, or spread the accusations online?
- Would my actions help to expose otherwise unseen sins that now actively threaten the well-being of others who are unaware?[26]
- Has my faultfinding made assumptions about another's motives, blinded me to God's grace at work in the person's life, and postured me in self-righteous smugness over him or her?
- If I do speak out, what is my presumed moment of resolution? Will my taking this action in public lead to an open-ended, unresolvable public dialogue that will inevitably grow into hostile irreconcilability and retaliation?
- Can I better serve the church by advocating key solutions and resolutions to a particular weakness emerging in the church (proactively) rather than addressing a person or a particular situation (retroactively)?

---

25. For a good summary of the complexities, see Baxter, *The Practical Works of the Rev. Richard Baxter*, 6:389–90.
26. Eph. 5:8–13.

The convenience of social media means I must be diligent to avoid overprioritizing the world's power structures,[27] careful not to ignore the supernatural power of two or three "called" Christians in a situation, and zealous to operate from pure motives. I must pray for God's help to be peaceable, gentle, open to reason, eager to offer mercy, and impartial in every complex situation.[28]

Social media can be used to confront major sin patterns and public heresies, yes. But when it comes to the dirt we have on one another, we must walk with the greatest care. Christians, of all people, should be most vigilant not to unnecessarily shovel one another's dirt into public view.[29]

ALWAYS REFORMING

I hope the digital sphere will become a more humane place over time, but I am certain the sin of backbiting will not disappear any time soon—it's too woven into the sinner's flesh, too embedded in the sinner's quick-to-slander heart. We must learn to distrust our sinful gut reactions and respect the institutions God has set in place in the church and, when necessary, civil law enforcement.

As we step back for an honest look at our digital era, we realize that our smartphones and social media help feed our generation's outrage. Most of us know firsthand what it's like to participate in slander. The most viral emotion is anger; the most viral story is scandal.

Since God put a biblical process in place to process accusations of sin in the lives of pastors,[30] I believe we should not be surprised when occasional scandalous pastors or situations emerge in our churches. There's a process in place for this, and I hope it is not often needed for pastors, certainly less than one in twelve (the failure rate of the original apostles). Whatever the percentage, we should be saddened, but not surprised. Churches and church leaders will sin seriously at times, and when they do, the important work of gathering facts,

---

27. 1 Cor. 1:18–31.
28. James 3:17.
29. 1 Cor. 6:1–8.
30. 1 Tim. 5:17–21.

dispelling myths, adjudicating accusations, confronting sinners, and caring for victims is too important, too complex, and too sensitive to be rendered "convenient" by the techniques of social media. But neither should the need for this work surprise us. We need one another to help shine the light of truth on our internal weaknesses, to help us repent of our sins against one another, and to pray with us for God's grace in our unending pursuit of maturity through reproof. *Ecclesia reformata, semper reformanda*—the Reformed church is always in need of reforming. And this important self-rebuke happens as the church continues her forward momentum of gospel mission into the world.[31]

OPTIMISM IN THE AGE OF OUTRAGE PORN

James 4 and command nine rebuke our appetite for "outrage porn"—a cultural appetite fed by click-seekers who pander to our "impulses to judge and punish and get us all riled up with righteous indignation."[32] Instead, God has written a script to help us honor, love, and care for one another—because we are sinners who fail one another and who need one another. Humbled low in the awesome presence of God, of Christ Jesus, and of the elect angels, we are charged to stop ourselves from making "prejudgments" about the moral condition of one another—judgments that are often motivated more by our personal prejudices and our easily provoked spirits of partisanship than anything else.[33]

In situations where we are not called to intervene, we are silent. In situations where we are called, we speak and confront in order to foster repentance in private. In all situations, at all times, as representatives of Christ, we are eager to resolve conflicts and be peacemakers. We aim to "outdo one another in showing honor" (Rom. 12:10). When we find ourselves insulted, we bless; when slandered, we entreat; when verbally persecuted, we endure.[34] At all costs, we

---

31. Karl Barth, Geoffrey William Bromiley, and Thomas F. Torrance, *Church Dogmatics*, vol. 4, part 3.2, *The Doctrine of Reconciliation* (London; New York: T&T Clark, 2004), 779–80.

32. Tim Kreider, "Isn't It Outrageous?" *The New York Times* (July 14, 2009). The phrase "outrage porn" was coined here.

33. 1 Tim. 5:21.

34. 1 Cor. 4:12–13.

do not become *irreconcilable*. We do not become men or women who ignite controversies in the church with no intention of pursuing healing and timely reconciliation.[35]

In this rough-and-tumble world, Paul and Silas model Christ-centered optimism. They were slandered with charges meant to destroy their reputations and were hammered with physical beatings intended to batter their bodies. Yet sitting in a prison cell in the midnight darkness and throbbing pain, they were found praying and singing hymns to God.[36]

"Our culture is looking for something to be angry, frustrated, and outraged about," pastor Matt Chandler said about Facebook discussions. "We thrive on pessimism. We want to be acutely aware of the brokenness of things and others, and that reveals something about us—God help us. But in light of this, should not Christians be annoyingly optimistic? We mourn with those who mourn. We weep with those who weep. We are a people who are easily heartbroken. But not easy to whip into a frenzy." This is true because "our God has never panicked."[37] He is in sovereign control.

We may live in an age of "outrage porn," but as children of the sovereign King, who has already won the climactic victory in the universe, we have no cause for pessimism. We have every reason to joyfully and optimistically "stick to the script."

---

35. Irreconcilability (ἄσπονδος, *aspondos*) is a sin that will heighten in the last days, says Paul (2 Tim. 3:3). It emerges in one who is unwilling to reconcile, unwilling to be at peace with others, and unwilling to negotiate a solution to a problem involving a second party. He is implacable— "unappeasable" (ESV), "irreconcilable" (NASB), and "unforgiving" (NIV). Paul chooses an ancient Greek war term meaning, at its root, refusal to enter into a treaty, to take a posture in which no flag of truce is allowed to pass between the parties, no terms of reconcilement are listened to. The irreconcilable party refuses to bring a state of war to an equitable close. Even in a stalemate, he will not lay down his weapon. As Paul makes clear to pastor Timothy, this person will not only keep up the fight, he will also contend he is acting in accord with the Christian faith, maintaining that his irreconcilability is biblically justifiable. It is not, and so we must avoid him—or block him, mute him, or do whatever is necessary to avoid him online (v. 5).

36. Acts 16:16–25.

37. Matt Chandler, sermon, "Who Was Conceived by the Holy Spirit, Born of the Virgin Mary," The Village Church, thevillagechurch.net (Sept. 13, 2015).

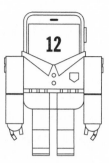

WE LOSE OUR PLACE IN TIME

Whether we realize it or not, every waking moment of our lives, we are asking ourselves questions: What should I do? What should I say? What should I stop? What should I start? We exist in time and space, and the priceless moment before us is ours to embrace. Right now, you are reading this book (for which I am grateful), but I am no longer writing this book. As I wrote this sentence, you were probably not aware that I was in the process of writing. My life decisions in the past and your life decisions in the present converge in this sentence. Lives intersect like this, in moment-by-moment investments.

We are all creatures made by God, eternal beings with no end to our conscious existences. You and I exist endlessly, called to forever make God look as satisfying as he really is (see chapter 6). That means we have been given the gift of this moment for faith, obedience, and trust in Christ.

Yet we live in a technologically driven culture, and we are conditioned to reflexively respond to "breaking news" in our feeds rather than to reflectively connect our past with our endless future (see chapter 9). This miscalibrated time stamp introduces the final way our smartphones are changing us. Like trying to focus on flashes of images as we scroll our social-media feeds, we microtask the fragments of life: a new fragment in an email discussion, a new fragment in a text conversation, a new fragment in a Twitter dialogue. In chasing after all these new fragments, we simply lose our place in time.

A TIME TO TWEET

No generation in the history of the world has been more capable of welcoming distractions into daily life, more likely to be pulled in various directions, and more prone to communicating in multiple simultaneous conversations. Twitter's 140-character mechanism for sharing brief thoughts has become a cultural metaphor for how amplified this effect has become in the digital age. In the spirit of Ecclesiastes 3:1–8, for every season there's a tweet:

A tweet to announce births,
and a tweet to announce deaths.
A tweet to criticize,
and a tweet to encourage.
A tweet to weep,
and a tweet to laugh.
A tweet to mourn,
and a tweet to dance.
A tweet to embrace,
and a tweet to repel.
A tweet to tear,
and a tweet to mend.
A tweet for war,
and a tweet for peace.

The pointless, the charming, the desirable—all these seasons now stack tightly together in linear tabulation in one vertical feed with no end. Compressed to fit into our phones, all these tweets hit us in one scrolling chronology. At one moment, we are called to weep with those who weep, and in another to rejoice with those who rejoice.

Life online is a whiplash between deep sorrow, unexpected joy, cheap laughs, profound thoughts, and dumb memes. Our social-media feeds give us what is sometimes riotous, sometimes amazing, sometimes dizzying, and sometimes depressing. But the disjointedness is something we have welcomed on ourselves.

King Solomon keenly observed that our souls must handle being tossed back and forth, because life is a series of changing seasons that require changing responses. And God created us to carry multiple emotions simultaneously, such as joy and sorrow.[1]

But in the digital age, those seasons come at us too quickly, and because they hit and leave so soon, we seldom feel the weight of our emotions. Behind the safety of our phone screens, we can more easily shield ourselves "from direct contact with the pain, the fears and the joys of others, and the complexity of their personal experiences." This doesn't make us suppress emotion; it makes us express "contrived emotion."[2]

We grow emotionally distant with our expressions. We become content to "LOL" with our thumbs or to cry emoticon tears to express our sorrow because we cannot (and will not) take the time to genuinely invest ourselves in real tears of sorrow. We use our phones to multitask our emotions. In the age of the smartphone, we are both trying to escape emotion and trying to "plug the need for contact with the drug of perpetual attention."[3] This juxtaposition, by necessity, makes us broadly connected but emotionally shallow.

LOSING TRACK OF TIME

At root here is a fundamental assumption about how many tweets and personal updates and news feeds, with their fractured patterns, are healthy for us. If we shortcut our emotional responses, and if we refuse to slow our lives to feel proper emotions, then there is one uncomfortable question a Christian must ask in an entertainment-driven culture, a question that never leaves me feeling more affirmed after asking it: Am I *entitled* to feed on the fragmented trivialities

---

1. 2 Cor. 6:10.
2. Pope Francis, "Encyclical Letter, Laudato Si' of the Holy Father Francis on Care for Our Common Home," The Holy See, w2.vatican.va (May 24, 2015).
3. Olivia Laing, *The Lonely City: Adventures in the Art of Being Alone* (New York: Picador, 2016), 247.

online? In other words, am I *entitled* to spend hours every month simply browsing odd curiosities?

I get the distinct sense in Scripture that the answer is *no*. I am not my own. I am owned by my Lord. I have been bought with a price, which means I must glorify Christ with my thumbs, my ears, my eyes, and my time.[4] And that leads me to my point: I do not have "time to kill"—I have time to redeem.

Yet smartphone abuse causes us to squander precious hours and almost erases us from our place in time in three different ways.

First, and most commonly, we simply lose track of time. Rapper and pastor Trip Lee told me: "I will admit, there have been times when I have looked up and realized I was looking down at my phone for fifteen minutes and my son was playing right in front of me, or I realized that I was not paying attention to my wife like I should. It takes intentionality, and that is an ongoing fight for me."[5] We get lost in the virtual world and become oblivious to the flesh-and-blood world around us, and we lose our sense of time.

Second, it is of the nature of technology to dislocate us historically. In principle, writes Craig Gay, "the technological habit of mind is *anti-teleological*. It is largely uninterested, and indeed incapable, of appreciating the notions of final causality or ultimate purpose."[6] Our digital devices cannot lead us, they cannot map our history, they cannot settle our priorities—all of these aims are rendered insignificant in comparison to the *now* of innovation.

Third, and most significant, if we use our phones to find sin, we cut ourselves off from God's timeline. In the Bible, the destructiveness of idolatry is nowhere more prominent than when it comes to *remembering*, and this is because cultural idols are the most poignant expression of God-forgetting.[7] Idols cut us off from remembering the

---

4. 1 Cor. 6:19–20.
5. Trip Lee, interview with the author via Skype (March 25, 2015).
6. Craig M. Gay, *The Way of the (Modern) World: Or, Why It's Tempting to Live as If God Doesn't Exist* (Grand Rapids, MI: Eerdmans, 1998), 92, emphasis original.
7. Ex. 20:22–24; Ps. 135:13–15; Isa. 44:19–22; 46:6–9; 57:11–13; Jer. 14:21–22; Ezek. 16:20–22; Jonah 2:7–8; 1 Pet. 4:1–6.

past mercy of God and blind us to his future grace. Idolatry skews the whole way you see yourself inside of the story written by the Creator.

Digital pornography is one specific example of how sexual fragmentation—a discrete moment of idolatrous lust—erodes our human identity and disconnects us from history. "Yes, pornography objectifies and commodifies the human body," explains historian Carl Trueman. "Yes, it alters neural pathways. Yes, it hinders healthy relationships. But it also cultivates an understanding of the human self which is profoundly disconnected from historical context, from the cosmic to the individual and all points in between."[8] Human sexuality is a created reality, designed by God, meant to weave together the fabric of human existence, generating new family units and producing future generations. Pornography rips sexuality out of this creational context and historical significance.

All of this forgetting and fragmenting is why we must never stop returning to our identity in Christ. In him, the powers of sin have been broken. We are no longer bound to obey our eye lust, bound to seek the approval of man, bound to find our relevance in viral memes, or addicted to what's trending on Reddit. My appetite for diversions and new daily curiosities has been crucified with Christ, and it is no longer the old me that lives online, but Christ living in me, and the life I now live online I live by faith in Christ, who loved me so much that he shed his blood for me.[9] All of this has a historical point, because in Christ I have a past, a present, and a future, and I now find my identity as one "on whom the end of the ages has come" (1 Cor. 10:11).

RUN!

The sun is the center of our physical solar system, but the earth is the center of the spiritual cosmos—which means that to be a human, to be a moral being alive in time and space, is to exist spiritually in the center phase of the most important race happening right now. We dare not slack off into the shadows of lethargy. All eyes are fixed on

8. Carl R. Trueman, "Sex Trumps History," *First Things* (March 15, 2016).
9. Gal. 2:20.

you and me. The spiritual adrenaline is pumping. Forget for a moment your virtual crowd of online followers and imagine all of your spiritual ancestors in the faith watching in the bleachers. Their times are legend; your time is now. Whether you were expecting it or not, the baton of faith, passed down from generation to generation, has now been slapped into your hands.[10]

Run!

Run with diligence. Cast off everything that distracts, unfetter your life from the chains that trip your ankles, and bolt with freedom and joy as you follow Christ. It is here, now, that the Spirit works tirelessly. It is here, now, that the work of Christ proves triumphant in the world. It is here, now, that the powers and principalities, defeated at Calvary, are being flaunted in defeat by the unity of the church.[11] The race is on—our race! We have one shot, one event—one life. We must shake off every sinful habit and every ounce of unnecessary distraction. We must run.

DIGITAL CHITCHAT

Redeeming the time—and recapturing a sense of our place in time— finally raises the question of digital chitchat. If we are honest, we use most of the time we spend on our phones for sharing jokes, GIFs, images, and videos, and for talking about sports, the weather, humor, and entertainment with our friends and family members. Digital "small talk" is a common use of our phones, and it's important for us to think through it carefully.

A growing example of digital chitchat comes in the aptly named platform Snapchat. This platform shatters the stereotype that social media is a place to collect a growing scrapbook of overly edited images from our lives. Instead, Snapchat is designed to give a place to more raw, open, honest, unfiltered, and unedited photos. Built for sharing "instant expressions" and "throwaway selfies," Snapchat allows users to instantly capture a moment in life with a camera. The image or

---

10. Heb. 12:1–2.
11. Eph. 3:7–4:16.

video is shared and, once opened, exists for only a few seconds. The app was designed to discretize a moment in time, separate it from a user's broader online history, and dislocate it from any larger life context so that it can be shared and soon deleted forever.[12]

Snapchat amplifies our discussion of digital chitchat, but it also helps us see the power of our phones to give us convenient touch points with our friends and family members, a power that is beyond question. It's nice to check in with others with a small glimpse into our lives or with a little humor, and our phones make this incredibly convenient.

An app such as Snapchat is vital to understanding our smartphones, writes Alastair Roberts. "While some might have expected the Internet and mobile phones chiefly to be used for the communication of *information*, their primary significance in most people's lives is their provision for the communication of *presence*. The Internet often feels a lot less like an 'information superhighway' and much more like a virtual village, where, through countless intertwined lines of relationship, everyone is minding everyone else's business."[13]

That's true. And added to this modern phenomenon is Jesus's ancient warning about the words we speak to one another each day: "I tell you, on the day of judgment people will give account *for every careless word they speak*, for by your words you will be justified, and by your words you will be condemned" (Matt. 12:36–37).

In this context, a "careless word" is literally a word "uttered without any thought of the effect it will have on other people."[14] We must be willing to put a stop to our lazy, *thoughtless* digital chitchat texts, humorous tweets, and laughable Facebook posts. But what if our smartphone chitchat has a purpose?

C. S. Lewis was terrifyingly right in his warning that our words push one another along one of two eternal trajectories (chapter 5).

---

12. As explained by Snapchat CEO Evan Spiegel, "What Is Snapchat?" YouTube, youtube.com (June 16, 2015).

13. Alastair Roberts, "Twitter Is Like Elizabeth Bennet's Meryton," *Mere Orthodoxy*, mereorthodoxy.com (Aug. 18, 2015), emphases original.

14. Leon Morris, *The Gospel according to Matthew*, The Pillar New Testament Commentary (Grand Rapids, MI: Eerdmans, 1992), 322.

And he anticipates the question I've asked. Does Jesus's warning require us to limit our conversation to relatively brief interactions when we must pass along essential information? Is there any room for us to have fun with one another?

Yes! "We must play," Lewis says about our relationships. "But our merriment must be of that kind (and it is, in fact, the merriest kind) which exists between people who have, from the outset, taken each other seriously—no flippancy, no superiority, no presumption."[15] We never jest about sin, and we don't joke to puff ourselves up or to harm others.

We can take Lewis's advice a step further here and see that digital joking and chitchat with one another can be done with an ultimate purpose. "Jesus does say every word counts," says Bible counselor David Powlison on the broader theme of small talk. "Even if we're just casually chatting, at heart that conversation is either a way for me to keep you at a distance, or a way to build a bridge between us. Small talk can be saying: 'I don't want to know you, and I don't want you to know me,' so I'm going to keep it light, as quick as possible, and see you later. Or, small talk can be a way to say, 'I care about you and I'd like to get to know you.' We might start by talking about football, or the weather—but it's heading somewhere more honest," he says. "Our small talk is going to be judged by God for its deeper intentionality."[16]

Powlison's wisdom here is key: God will judge our digital conversations, private texts, and public tweets by the intentions of our hearts. So I ask myself: Is my digital chitchat aimed or is it aimless; thoughtful or thoughtless; strategic for the eternal good of others or wasted on self-expression? Has my digital chitchat habituated all of my conversations online, reducing my words to nothing more than slapstick fun?

If I consider my phone only as a tool to "instantly express" my life, then my phone use is vain. I must ask: Am I lazy and careless with

15. C. S. Lewis, *The Weight of Glory: And Other Addresses* (New York: HarperOne, 2001), 46.
16. David Powlison, email to the author (May 13, 2016). Shared with permission.

souls, ignorant of the power of words, images, and links on others? Or am I using my digital chitchat as a way to build into someone (or some online community) with a larger relational goal of edification? These questions determine whether my texts, tweets, and images are thoughtless fragments or purposeful strategies to point others to find their joy, meaning, and purpose in God. This is digital chitchat with historical (and eternal!) purpose.

A THEOLOGY OF REMEMBERING (AND FORGETTING) IN THE PSALMS

We are talking about time, history, and finding our joy, meaning, and purpose in God—and helping others toward these goals is the aim not only of the Christian life, but of one very important spiritual discipline.

The Psalms have more to teach than any other book in the Bible about the spiritual discipline of *remembering* (and the spiritual dangers of *forgetting*). Psalms 42 and 77 are like lights to illuminate God's past kindnesses when our present realities seem dark. When pain enters our lives and we feel the sting of this world, we draw on God's fidelity. Psalm 78 is a plea that every coming generation of believers will be taught to remember God's goodness, but it is also a warning that they must not become so caught up in the distractions of life that they forget the mighty works of God. Remembering is also the theme of the more upbeat Psalm 105.

On the flip side, God's people plead with him to remember them in Psalm 74. Psalm 9 assures us that he won't forget us.

Throughout the Old Testament, believers find strength and safety merely in the act of remembering God. "Some trust in chariots, and some in horses: but we will *remember* the name of the LORD our God" (Ps. 20:7 KJV).

It is a spiritual discipline to *remember* God's acts of deliverance.

> Bless the LORD, O my soul,
>   and all that is within me,
>   bless his holy name!

Bless the LORD, O my soul,
  and forget not all his benefits,
who forgives all your iniquity,
  who heals all your diseases,
who redeems your life from the pit,
  who crowns you with steadfast love and mercy,
who satisfies you with good
  so that your youth is renewed like the eagle's. (Ps. 103:1–5)

We will not neglect God's precious Word, because we delight in it and cherish it.[17] We remember God's mighty works of old like a well-ingrained habit, and this discipline stokes our souls' desire to taste more of the precious beauty of God.[18]

To remember God is to satisfy the soul and to recalibrate our always-shifting perception of reality. But to forget God is to forsake God. This spiritual plague of forgetfulness is not physical forgetfulness or mental dementia. Spiritual forgetting is sin, a sin that plagues youth[19] and infests every demographic.

A THEOLOGY OF REMEMBERING (AND FORGETTING) IN THE NEW TESTAMENT

The discipline of *remembering* carries over into the New Testament. Every dimension of the Christian life seems to be defined in some way by the command to remember. For example:

- We remember Christ's body and blood as we break bread and drink from the cup in the Lord's Supper, doing all of this often, "in remembrance" of him.
- We remember the all-powerfulness, all-goodness, and all-presence of Jesus Christ, which is essential for our fulfillment of the Great Commission.
- We remember the history of our dark sins so that the beauty of Christ's grace shines on our present (and future) forgiveness.

---

17. Ps. 119:16.
18. Ps. 143:5–6.
19. Eccles. 12:1–8.

- We remember Lot's wife, and we turn our eyes away from the worthless idols of this life.
- We remember the long history of persecution of God's people to be reminded that the tensions we feel in our own culture are not strange.
- We remember the grace at work in our closest brothers and sisters in order to thank God for them.
- We remember the needs of our immediate brothers and sisters in order to genuinely pray for them.
- We remember the physical needs of our remotest Christian brothers and sisters on earth so that we can care for them from a distance.
- We remember that God disciplines us because he loves us, and so that we can grow in grace, humility, and joy.

Scripture tells us that God is not unjust so as to forget our works and our demonstration of love in serving the saints,[20] even as, in Christ, he remembers our sins no more.[21]

What is true in the Old Testament is also true in the New Testament. God wants us to *remember* the script he has written for our lives, especially his acts of redemption.[22] This is because Christ inaugurated time,[23] he now upholds time,[24] and he has the sovereign power to unroll the events that will end time.[25] At every moment of history, Christ speaks of himself: "I am the Alpha and the Omega, the first and the last, the beginning and the end" (Rev. 22:13). He is the keeper of all time and history, and he alone holds my forever-history secure.[26]

NEVER FORGETTING

Whatever else is at play in the digital age, Christians are commanded over and over to *remember*. We must not lose our past and our future

---

20. Heb. 6:10.
21. Heb. 8:12; 10:17.
22. Eph. 2:11–13.
23. Rev. 4:11.
24. Heb. 1:3.
25. Revelation 5–6.
26. Jude 24–25.

for moment-by-moment tweets and texts on our phones. But our remembering is not like flipping through a dusty scrapbook of reminiscence. The Bible cuts into our hearts with a living-and-active memory for daily life in the digital age. The Word calls us to remember in order to obey, as the apostle Peter explained when he said that our aim is Christian maturity that grows from faith to goodness, from knowledge to self-control, from perseverance to godliness, and finally from mutual affection to love. "For whoever lacks these qualities is so nearsighted that he is blind, *having forgotten* that he was cleansed from his former sins" (2 Pet. 1:9). All spiritual growth is rooted in remembering what Christ has done in me.

Remembering is a key verb of the Christian life. We recall our past, we correct our nearsightedness, we take heart, we regain mental strength, we find peace in the eternal Word. Remembering is one of the key spiritual disciplines we must guard with vigilance amid the mind-fragmenting and past-forgetting temptations of the digital age.

Conclusion

LIVING SMARTPHONE SMART

In the last twelve chapters, I have warned against twelve corresponding ways in which our smartphones are changing us and undermining our spiritual health:

- Our phones amplify our addiction to distractions (chapter 1) and thereby splinter our perception of our place in time (12).
- Our phones push us to evade the limits of embodiment (2) and thereby cause us to treat one another harshly (11).
- Our phones feed our craving for immediate approval (3) and promise to hedge against our fear of missing out (10).
- Our phones undermine key literary skills (4) and, because of our lack of discipline, make it increasingly difficult for us to identify ultimate meaning (9).
- Our phones offer us a buffet of produced media (5) and tempt us to indulge in visual vices (8).
- Our phones overtake and distort our identity (6) and tempt us toward unhealthy isolation and loneliness (7).

But it's not just about warnings. Along the way, I have also attempted to commend twelve life disciplines we need to preserve our spiritual health in the smartphone age:

- We minimize unnecessary distractions in life to hear from God (chapter 1) and to find our place in God's unfolding history (12).
- We embrace our flesh-and-blood embodiment (2) and handle one another with grace and gentleness (11).
- We aim at God's ultimate approval (3) and find that, in Christ, we have no ultimate regrets to fear (10).
- We treasure the gift of literacy (4) and prioritize God's Word (9).
- We listen to God's voice in creation (5) and find a fountain of delight in the unseen Christ (8).
- We treasure Christ to be molded into his image (6) and seek to serve the legitimate needs of our neighbors (7).

The book is organized into a chiasm so that everything centers inward on chapters 6 and 7, which focus on the two greatest commandments that frame our identity and define our purpose on earth: love God (6) and love your neighbor (7). Scripture makes life focus possible in the digital age, and it does so when Jesus boils down the purpose and aim of our lives into two goals: treasure God with your whole being, and then pour out your God-centered joy in love for others.[1] On these two commands all other smartphone laws depend.

SATAN'S "NOTHING" STRATEGY

At some point, we must leave the pages of this book to struggle with these laws in the real world.

Today, tired after work, I opened Facebook on my phone, looking for a diversion. I flicked past a video of a cat that sounds like a crying child; then I saw a new study about gun control; then I saw an innovative new keyboard for tablets; then I read a story from the latest celebrity gossip; then I was offered twenty pictures of actors who have aged badly (which I ignored); then I saw a breaking news story about a rogue militia group in Oregon; then I read that North Korea

---

1. Matt. 22:34–40.

apparently had detonated a test atomic bomb; then I watched a viral video of a "monster shredder" that crushes refrigerators, couches, and cars with large metal teeth; and then I saw pictures of a friend and his wife on vacation in Iceland. On and on I flicked down a list of disconnected and fragmented items, and most of them only barely important or interesting. I was not edified or served, only further fatigued because of missing a nap I should have taken or a walk I could have taken, and easily lured back to my phone for more. And then I remembered I skipped my personal disciplines this morning. My battle against all the slothful smartphone tendencies I see in my own heart has only begun.

What I am coming to understand is that this impulse to pull the lever of a random slot machine of viral content is the age-old tactic of Satan. C. S. Lewis called it the "Nothing" strategy in his *Screwtape Letters*. It is the strategy that eventually leaves a man at the end of his life looking back in lament: "I now see that I spent most of my life in doing neither what I ought nor what I liked."[2]

This "Nothing" strategy is "very strong: strong enough to steal away a man's best years, not in sweet sins, but in a dreary flickering of the mind over it knows not what and knows not why, in the gratification of curiosities so feeble that the man is only half aware of them . . . or in the long, dim labyrinth of reveries that have not even lust or ambition to give them a relish, but which, once chance association has started them, the creature is too weak and fuddled to shake off."[3]

Routines of nothingness. Habits unnecessary to our calling. A hamster wheel of what will never satisfy our souls. Lewis's warning about the "dreary flickering" in front of our eyes is a loud prophetic alarm to the digital age. We are always busy, but always distracted—diabolically lured away from what is truly essential and truly gratifying. Led by our unchecked digital appetites, we manage to transgress both commands that promise to bring focus to our lives. We fail to enjoy God. We fail to love our neighbor.

---

2. C. S. Lewis, *The Screwtape Letters* (New York: HarperOne, 2001), 60.
3. Ibid.

Amid these habits of nothingness, we find ourselves wandering half-awake in digital idleness, prone to leave our digital responsibilities to become digital busybodies and digital meddlers.[4] We give our time to what is not explicitly sinful, but also to what cannot give us joy or prepare us for self-sacrifice. Satan's "Nothing" strategy aims at feeding us endlessly scrolling words, images, and videos that dull our affections—instead of invigorating our joy and preparing us to give ourselves in love.[5]

IDOL?

Technology makes life easier, but immaturity makes technology self-destructive. With my phone, I find myself always teetering between useful efficiency and meaningless habit. I am often reminded that my phone may be a lot of things, but it is not a toy. The magician and the wielder of a smartphone are close cousins,[6] and this is because, suggests literary critic Alan Jacobs, our modern technology offers us a bewitching power not unlike the magic in the Harry Potter fantasy series: "Often fun, often surprising and exciting, but also always potentially dangerous. . . . The technocrats of this world hold in their hands powers almost infinitely greater than those of Albus Dumbledore and Voldemort."[7] Into our hands are placed these wands, these smartphones, these powers of idolatry, freighted with redemptive expectations.

The digital age can bewitch and capture our hearts in unhealthy ways. Our advances in technology have a way of rendering God more and more irrelevant to our world and in our lives—the very definition of worldliness.[8] And if our digital technology becomes our god, our wand of power, it will inevitably shape us into technicians who

---

4. 1 Thess. 4:11; 2 Thess. 3:11; 1 Tim. 5:13; 1 Pet. 4:15.

5. This same principle is explained well by Puritan Richard Baxter in *The Practical Works of the Rev. Richard Baxter* (London: James Duncan, 1830), 3:535–36.

6. C. S. Lewis, *The Abolition of Man or Reflections on Education with Special Reference to the Teaching of English in the Upper Forms of Schools* (New York: HarperOne, 2001), 76–77.

7. Alan Jacobs, *A Visit to Vanity Fair: Moral Essays on the Present Age* (Grand Rapids, MI: Brazos, 2001), 147–48.

8. See Craig M. Gay, *The Way of the (Modern) World: Or, Why It's Tempting to Live as If God Doesn't Exist* (Grand Rapids, MI: Eerdmans, 1998).

gain mastery over a dead world of conveniences. Aimlessly flicking through feeds and images for hours, we feel that we are in control of our devices, when we are really puppets being controlled by a lucrative industry.

While our techniques of control do not make us atheists, they do seem to make worship more and more irrelevant, as God is more and more displaced from our lives. We forget how to meet God, and yet we defend our smartphones, unwilling to admit that we are more concerned with controlling the mechanics of our lives than in worshiping the God whose sovereign power directs our every breath.

We must watch for signs that our worship is veering off course. We can no longer simply worship God in admiration or pray to him without a compulsive fidgeting for our phones. We talk more about God than we talk to him. Our hearts are more interested in following empty patterns of worship than encountering the Spirit. Our worship on Sunday seems flat, but our week is filled with an endless quest for Christian advice to fix what we know is wrong. We seek a mechanical relationship with God, searching for new techniques to fill the spiritual void in our lives. Signs such as these reveal how technology degrades our priorities. But worship calls for redirection in our lives.

TECHNIQUES OF HOLINESS?

"A hundred years ago no Christian would ever have thought of writing a book called *Three Easy Steps to Being Filled with the Spirit*," said pastor Tim Keller. "You see, on the one hand, we've been so affected by our technological society that we want to make everything a commodity. Let's boil everything down to procedures. I want to be in control."[9] That's how life now works.

Keller raises a critical warning sign for Christian pilgrims in the digital age. In our love of mechanisms, techniques, and power, we lose our way—and we lose our worship and our prayer, because God

---

9. Timothy Keller, sermon, "Be Filled with the Spirit—Part 1," Gospel in Life, gospelinlife.com (June 16, 1991).

has grown secondary to our technology. But God is the sovereign King who will not bow to our gadget mastery. Apps can help me stay focused on my Bible reading plans and help me organize my prayer life, but no app can breathe life into my communion with God.

Self-criticism in the digital age is a necessary discipline—an act of courage. "It is by being able to criticize that we show our freedom. This is the only freedom that we still have, if we have at least the courage to grasp it."[10] Our personal freedom from the misuse of technology is measured by our ability to thoughtfully criticize it and to limit what we expect it to do in our lives. Our bondage to technology is measured by our inability to thoughtfully criticize ourselves. What shall it profit a man if he gains all the latest digital devices and all of the techniques of touch-screen mastery but loses his own soul?

Are we courageous enough to ask?

In truth, the automated voice inside of my smartphone can find a local restaurant for me or tell me when to leave in order to beat the traffic. But my phone can never fulfill my greatest needs in life. My phone (like any technology) cannot explain why I exist, cannot define the end and aim of my life, cannot tell me if I've lost my way, cannot order my life priorities, and cannot tell me what choices in life are morally right or wrong.

In an act of courageous self-criticism, I must ask three questions:

- Ends: Do my smartphone behaviors move me toward God or away from him?
- Influence: Do my smartphone behaviors edify me and others, or do they build nothing of lasting value?
- Servitude: Do my smartphone behaviors expose my freedom in Christ or my bondage to technique?

HUMBLE LISTENING

So, should Christians trade their smartphones for dumbphones? This decision must be made by each of us as we listen to the leading

---

10. Jacques Ellul, *The Technological Bluff* (Grand Rapids, MI: Eerdmans, 1990), 411.

of the Holy Spirit in our lives. We pay more attention to our phones than we do to the third person of the Trinity, but he cares for us more than we care for ourselves. Perhaps you believe you would benefit spiritually by stepping away from your phone for a season. Or perhaps you feel led to rethink better boundaries in your digital life. Or you may be fed up with your love-hate-deactivate-delete-reactivate relationship with social media, and you are ready to rid yourself of your smartphone altogether. I cannot tell you what to do, but I can encourage you to heed the conviction of the Spirit, who will help you make the next step of obedience.

Some smartphone users prove adept at balancing smartphone use and not falling into the powerful lures and traps explained in this book. Some blend the strengths of digital blessings into their lives, so that they resemble healthy digital centaurs. But not all of us can carry this balance.

For all of us, the challenge is in extending grace to one another. Technophobic pride says, "God, I thank you that I'm not like this gadget addict who is distracted by his devices and feeding on the banal trivialities of the fake world." Technophiliac pride says, "God, I thank you that I'm not like this tech despiser who is too undisciplined to manage the digital distractions of the real world." Both views are arrogant.

SMARTPHONE DIVERSITY

The church needs Christians who use, and don't use, smartphones. As I said in the preface of this book, smartphone habits expose the heart, which means that the solution to unhealthy smartphone habits is not found merely in the embrace of the predigital utopia of typewriters and vinyl records. Simply calling all Christians to ditch their smartphones is no magical solution, because without genuine humility, true confession of sin, and supernatural heart change, we will not be free from the banal distractions and endless cotton candy allurements *offline* (see this story[11]). As Christians convinced by the

---

11. Paul Miller, "I'm Still Here: Back Online after a Year without the Internet," *The Verge*, theverge.com (May 1, 2013).

Spirit to take such a bold step, we must be humbled by God's grace and blessed with a vision for how new healthy priorities can replace our unhealthy habits (see this story[12]).

At the beginning of this project, theologian David Wells said that we cannot become digital monks. No, not all of us. Historian Bruce Hindmarsh was the first to suggest to me that the church needs a few young Christians who willingly live off the digital grid so that believers who are enmeshed inside the digital world can find a contrast and comparison for personal reflection. In his words, those off the grid function something like an astronaut living in outer space, who can return and report on what life is like in a different environment.[13] Or, if I can flip the metaphor (for those who object to the "digital monk" term[14]), we need people who live disconnected lives on earth so that we who are wired to the digital age, and now dangle in the outer space of technical innovation, can look back to see if our smartphones have really accelerated our lives—or if we are just floating aimlessly.

Either way, we need them—Christians who can, as much as possible, live offline (even as many predict the distinct terms *online* and *offline* will soon be things of the past).

For smartphone users, seasonal digital monkery will doubtlessly become an essential discipline for healthy Christian living. I could not have written this book without powering down my phone often. And when the writing process was especially intense, I turned off the WiFi to my computer. It felt very isolating at first, but over time, it became therapeutic and liberating as I pursued one of my seasonal callings.

We will benefit from returning often to the challenge of Francis Schaeffer, who said: "Christians have two boundary conditions: (1) what men *can* do, and (2) what men *should* do. Modern man does not have the latter boundary."[15]

---

12. Andrew Sherwood, "The Sweet Freedom of Ditching My Smartphone," *All Things for Good*, garrettkell.com (Jan. 21, 2016).

13. Bruce Hindmarsh, interview with the author via phone (March 12, 2015).

14. Alan Levinovitz, "I Don't Have a Cellphone. You Probably Don't Need One, Either," *Vox*, vox.com (March 15, 2016).

15. Francis A. Schaeffer, *The Complete Works of Francis A. Schaeffer: A Christian Worldview*, vol. 1, *A Christian View of Philosophy and Culture* (Westchester, IL: Crossway, 1982), 369, emphases original.

The essential question we must constantly ask ourselves in the quickly evolving age of digital technology is not what *can* I do with my phone, but what *should* I do with it? That answer, as we have seen, can be resolved only by understanding why we exist in the first place.

SHOULD I DITCH MY SMARTPHONE?

If there was a season in smartphone history when we faced the decision of whether or not to "opt in," that period was short and has passed. The only question now is whether we are going to "opt out." So we arrive at the gigantic question: Should I ditch my smartphone?

First, we must see that our phones are the aggregate of all our digital mechanisms, so much so that we often don't think about what we use and what we need on our phones. We can benefit from frequently disaggregating our smartphones, breaking down our technology in cost, features, and functions. For example, I often ask myself these twelve questions:

1. What does my smartphone cost me per year if I add up the price of the device, insurance protection, covers and cases, and monthly service?
2. Do I need mobile web access to fulfill my calling in vocation or ministry?
3. Is texting essential to my care for others? Do those texts need to be seen in real time? And is the smartphone the only way to do it?
4. Do I need mobile web access to legitimately serve others?
5. Do I need mobile web access to navigate unfamiliar cities? Is the device an essential part of my travels?
6. Do I need my smartphone to take advantage of coupons in stores? How much money would I save instead without a smartphone data plan?
7. Can my web access wait? Is the convenience of mobile web access something I can functionally replace with structured time at a laptop or desktop computer later?

8. Can I get along just as well with a dumbphone, a WiFi hotspot, an iPod, or a tablet?
9. Can I listen to audio and podcasts in other ways (through an iPod, for example)?
10. Have I simply grown addicted to my phone? If so, can the problem be solved with moderation, or do I need to just cut it off?
11. Do the mobile lures of my phone insulate me from people and real needs around me?
12. Do I want my kids to see me gazing at a handheld screen so much as they grow up? What does this habit project to them and to others around me?

These are significant questions.

Giving up a smartphone is not only one of the bravest and most countercultural acts of defiance possible today, it is a gift to others. If I'm a social-media junkie, my lack of self-control feeds the social-media addiction in you. And the more I text and tweet and Snapchat, the more I drag you and others into the digital vortex of reciprocating obligation. This is the secret to how social-media giants grow their valuations into the billions. They need *me* to entice *you*. Even something as simple as pulling out your smartphone in a crowd is "the new yawn"—everyone else around you will feel the immediate pull and lure to check their own phones.[16] Rarely do we think of how our own digital addictions impact others (especially our children), and rarely do we see this as one of the most daunting challenges in giving up the smartphone. To any addict brave enough to go smartphone free, I applaud you. You will serve the people around you in unseen ways that will never be noticed or celebrated.

LIVING SMARTPHONE SMART

For now, in this season of ministry, I will own a smartphone. But like never before, I can see how unnecessary the phone is to most of my life. I'm challenged to be far more disciplined than I ever imagined

---

16. Donna Freitas, *The Happiness Effect: How Social Media Is Driving a Generation to Appear Perfect at Any Cost* (New York: Oxford University Press, 2017), 218.

I would be. The writing of this book marks a new era in my relationship with digital technology.

Perhaps the clearest revelation of this project is simple: to benefit from my phone, I must not use all of the features all of the time. This is true because my phone is an open platform for developers to fill with shiny apps that promise me productivity or amusement. Contrary to Schaeffer's wisdom, we buy our phones with the unquestioned assumption that anything our devices *can do* they *should do*. Or, to say this more personally, we tend to fill our devices with a lot of nonessential apps. If this sounds weird, it is, because we have been conditioned to never ask the minimalist question: What is truly *essential* for my phone to accomplish?

We do ask this of other technology. Imagine me driving in my minivan. Based on the dashboard readout, my van can travel at 140 miles per hour (unconfirmed). So I *could* race the van every weekend on a local racetrack for fun. But that's not what the van is for. It's not intended to win races or to exceed speed limits. It exists to provide safe transportation for my family. To draw out the full benefit of my van, there is no need for me to use all the features at maximum capacity. If, in fact, my van *can* reach 140 mph (which I doubt!), that's so it can travel at 70 mph legally, safely, and comfortably. There are unsaid limits to what I ask the van to do. Certain features serve my family—others don't.

The key to balancing ourselves in the smartphone age is awareness. Digital technology is most useful to us when we limit its reach into our lives. The world will always expect technology to save humanity from its darkest fears, and to that end, it will submit more and more of itself to breaking innovations. But by avoiding the overreach of these misdirected longings for techno-redemption, we can simply embrace technology for what it is—an often helpful and functional tool to serve a legitimate need in our lives.

Every technology requires limits, and the smartphone is no exception. If you find the smartphone is absolutely necessary for your life and calling, put clear regulators in place. Consider these twelve boundaries:

1. Turn off all nonessential push notifications.
2. Delete expired, nonessential, and time-wasting apps.[17]
3. At night, keep your phone out of the bedroom.
4. Use a real alarm clock, not your phone alarm, to keep the phone out of your hands in the morning.
5. Guard your morning disciplines and evening sleep patterns by using phone settings to mute notifications between one hour before bedtime to a time when you can reasonably expect to be finished with personal disciplines in the morning (9 p.m. to 7 a.m. for me).
6. Use self-restricting apps to help limit your smartphone functions and the amount of time you invest in various platforms.
7. Recognize that much of what you respond to quickly can wait. Respond at a later, more convenient time.
8. Even if you need to *read* emails on your smartphone, use strategic points during the day to *respond* to emails at a computer (thirty minutes each at 9 a.m. and 4 p.m. for me).
9. Invite your spouse, your friends, and your family members to offer feedback on your phone habits (more than 70 percent of Christians in my survey said nobody else knew how much time they spent online).
10. When eating with your family members or friends, leave your phone out of sight.
11. When spending time with family members or friends, or when you are at church, leave your phone in a drawer or in your car, or simply power it off.
12. At strategic moments in life, digitally detox your life and recalibrate your ultimate priorities. Step away from social media for frequent strategic stoppages (each morning), digital Sabbaths (one day offline each week), and digital sabbaticals (two two-week stoppages each year).[18]

---

17. See the helpful app-management tips from Tristan Harris, "Distracted in 2016? Reboot Your Phone with Mindfulness," tristanharris.com (Jan. 27, 2016).

18. See Tony Reinke, "Know When to Walk Away: A Twelve-Step Digital Detox," Desiring God, desiringGod.org (May 30, 2016).

TEST YOURSELF

This book cannot end without considering the impact of our smartphones on the totality of our bodies. It is inexcusable that we fret more over charging our phones than we do over calculating how many hours of sleep our bodies need. We are embodied creatures, and that means that the way we use digital technology changes all of us—mentally, physically, and spiritually. Solomon warned us to not divorce our minds from our whole bodies, the very temptation of the touch-screen age.[19]

Study after study has shown that too much time on our phones has profound effects on our physical health, including (but not limited to) inactivity and obesity, stress and anxiety, sleeplessness and restlessness, bad posture and sore necks, eye strain and headaches, and hypertension and stress-induced shallow breathing patterns. The physical consequences of our unwise smartphone habits often go unnoticed, because in the matrix of the digital world, we simply lose a sense of our bodies, our posture, our breathing, and our heart rates.

Our overwhelming focus on projected images causes negligence with regard to our bodies. Go to YouTube, search for "texting and walking accidents," and you'll find a growing collection of video clips of smartphone users so engrossed with their phones that they unconsciously walk right into street traffic or walls, fall into public water fountains, or slip and get caught in sidewalk grates. Our phones have made us so physically oblivious to other people in public areas that "we have gone from holding the door out of courtesy to standing before it out of obliviousness."[20]

Failing to focus on the bodily consequences of our disembodied virtual habits is an oversight many are trying to correct.[21] One of my hopes for this book is a renewed self-awareness of how technology influences me—all of me. I want you to be aware of yourself, too.

---

19. Eccles. 12:12.

20. John Dickerson, "Left to Our Own Devices," *Slate*, slate.com (June 24, 2015).

21. See David M. Levy, *Mindful Tech: How to Bring Balance to Our Digital Lives* (New Haven, CT: Yale University Press, 2016).

But while I can outline some of your possible symptoms, I cannot diagnose you, and I certainly cannot tell you to get rid of your phone altogether.

You need to test your use of technology just as you would a physical diet. If you don't feel well after eating, you ask yourself whether it was because you overate, because you are allergic to what you consumed, or because you ate junk food, spoiled food, or poisoned food. Ask similar questions about your smartphone. What happens to your mind and body when you stay off Facebook for a week, when you don't answer emails remotely on your phone, when you don't sleep near your phone, or when you limit Twitter to certain times? And as you engage your phone, watch your breathing, your levels of anxiety, and your posture.

And do the same thing spiritually. Change your smartphone routines and see what happens to your devotional life. Are your mornings more fruitful and focused? What happens at church when you leave your phone in the car?

Listen to your body and listen to your soul, and use those evaluations to inform your smartphone habits. Use the negative impacts to evaluate your practices, and let the positive impacts inform your future strategies.

The questions we ask about our smartphones are urgent. Many of us would like to answer these questions with a list of smartphone rules, but we cannot simply copy and paste a single list into everyone's lives. As you determine your smartphone limits, use a rotation diet, pray, use your smartphone with God's wisdom, and by all necessary means, stay vigilant to avoid the trap of Satan's "Nothing" strategy.

EPILOGUE

At the end of this study, I find myself chastened for my seasons of iPhone abuse and motivated to set better parameters for myself (and for my family[1]). I am caught between anxieties about how my phone rewires my habits and impacts my body. I wonder if I can be reprogrammed, or if it's too late. And yet I marvel with gratitude over how my phone boosts my productivity and ministry reach.

Back when I connected with Oliver O'Donovan, the respected ethicist, I asked him: Should Christians feel uneasy about the rise of digital communications technology?

"Feeling uneasy is not a sufficient response," he wisely cautioned. "All that *can* be received from God with thanksgiving *should* be received with thanksgiving." That is a good pushback. He continued, "My generation was fifty, and very busy, when the first personal computers hit, and so we have probably never overcome our ambivalence at the sheer disruption and disturbance they caused as we had to relearn all our developed skills—*and then learn them again*, when the first wave of software gave way to the second." He said he has been forced to learn how to type the Greek alphabet in five or six different ways, with new software upgrades and changes over the years. Despite these drawbacks to technology, "I can still thank God for some things these innovations have given me, and I would wish my grandchildren to be able to thank God for more. But *how* to learn

---

1. See my family takeaways in Tony Reinke, "Walk the Worldwide Garden: Protecting Your Home in the Digital Age," Desiring God, desiringGod.org (May 14, 2016).

to thank God? One cannot thank God for anything that one cannot understand. It is a real and difficult question, and not just a matter of being upbeat and believing in progress."[2]

So how can we master our smartphones? How can we thank God for them while remaining prayerfully self-critical of our habits?

MODERN MARVELS

"I am just old enough to remember the world before telephones," wrote G. K. Chesterton (1874–1936) near the end of his life.[3] Chesterton was a toddler when Alexander Graham Bell secured a patent for a voice-replicating device, and by the time of Chesterton's death, phone calls were connecting across the Atlantic Ocean. Perhaps it is odd to end a book on the smartphone by looking back at the telephone, but since his life spanned both the invention and proliferation of this revolutionary device, Chesterton may have something to say to us.

The onslaught of digital technology in our day seems almost like magic, a kind of enchantment that should expand the awe and wonder of our souls. But typically it does not. The magic fizzles, warned Chesterton.[4] Rather, the technological revolution "has been a rapidity in things going stale; a rush downhill to the flat and dreary world of the prosaic; a haste of marvelous things to lose their marvelous character; a deluge of wonders to destroy wonder. This may be the improvement of machinery, but it cannot possibly be the improvement of man."[5]

Eventually, our technological wonders fail to win our admiration; they simply become more cold gears in the mechanized processes of daily living that we mindlessly master. The electronic starters on our car engines are a good example. Think of it: with nothing but

---

2. Oliver O'Donovan, interview with the author via email (Feb. 10, 2016).
3. G. K. Chesterton, *The Collected Works of G. K. Chesterton*, vol. 35, *The Illustrated London News: 1929–1931* (San Francisco: Ignatius, 1991), 252.
4. In the words of one modern-day novelist, "The sexier our high-tech stuff gets, the less I am able to feel anything about it." Charles Yu, "Happiness Is a Warm iPhone," *The New York Times*, nytimes.com (Feb. 22, 2014).
5. G. K. Chesterton, *The Collected Works of G. K. Chesterton*, vol. 37, *The Illustrated London News: 1935–1936* (San Francisco: Ignatius, 2012), 22–23.

the turn of a little metal magic wand, sparks ignite an explosive liquid refined from an ancient organic sludge—a sludge somehow sunk deep in the earth, covered over, liquefied by age and pressure to become a potion later sucked from subterranean cavities and processed into flammable fuel that is then pumped into tanks and finally into smooth cylinders carved from solid steel, where it meets those sparks—causing choreographed eruptions that pop so powerfully, so perfectly, and so consistently that we can, with fire and with one extended foot, thrust ourselves smoothly across the city on a magic carpet with four wheels.

However, when we twist the wand and the product of that ancient sludge erupts into a series of fireballs, we simply move along. Of course, if the battery dies or the clear brew is used up, and the magic wand proves powerless, we swear and fuss. But mostly, when everything works right, we don't notice.

OUR INDIFFERENCE TO WONDERS

We *must* notice these marvels, so in January 1935, Chesterton confronted technological amnesia in a column about modern marvels, titled "Our Indifference to Wonders."[6] It remains one of the most important contributions to Christians in the digital age. In his column, Chesterton labors to make his point with poignant paganism:

> Tell me that the bustling businessman is struck rigid in prayer at the mere sound of the telephone-bell, like the peasants of Millet at the Angelus; tell me that he bows in reverence as he approaches the shrine of the telephone-box; tell me even that he hails it with Pagan rather than with Christian ritual, that he gives his ear to the receiver as to an Oracle of Delphi, or thinks of the young lady on an office-stool at the Exchange as of a priestess seated upon a tripod in a distant temple; tell me even that he has an ordinary poetical appreciation of the idea of that human voice coming across hills and valleys—as much appreciation as men had about

---

6. Ibid., 21–24.

the horn of Roland or the shout of Achilles—tell me that these scenes of adoration or agitation are common in the commercial office on the receipt of a telephone call, and *then* (upon the preliminary presumption that I believe a word you say), *then* indeed I will follow your bustling businessman and your bold, scientific inventor to the conquest of new worlds and to the scaling of the stars. For then I shall know that they really do find what they want and understand what they find; I shall know that they do add new experiences to our life and new powers and passions to our souls; that they are like men finding new languages, or new arts, or new schools of architecture. But all they can say is that they can invent things which are generally commonplace conveniences, but very often commonplace inconveniences.[7]

In this excerpt, Chesterton feeds pagan mythology a dose of steroids. To put this in my own shorter translation, he seems to be saying that the premodern pagan was better suited for the technological age than the secular materialist is now. Chesterton was a prophet who saw the dawning of a disenchanted and mechanical world run by the techniques of technology. And he saw humans responding as something like affectionless robots. So in this column, he comes running at us with hands up, waving in protest. And his discord rings true today: "The modern system presupposes people who will take mechanism mechanically; not people who will take it mystically."[8] That was his fear. Chesterton believed that materialism was behind both ideas: the phone *will damn us* or the phone *will save us*. It is just as idolatrous to *blaspheme* a phone as it is to *worship* a phone. The solution is for us to wisely enjoy the smartphone—imaginatively, transcendentally, as something that should deepen our wonder.

So when one modern-day sociologist says that only those "capable of desacralizing technology can begin to search for meaning and

---

7. Ibid., 23–24, emphases original.

8. G. K. Chesterton, *The Collected Works of G. K. Chesterton*, vol. 5, *The Outline of Sanity; The End of the Armistice; Utopia of Usurers—and Others* (San Francisco: Ignatius, 1987), 152.

hope elsewhere," he is right, in a sense.[9] Our ultimate redemptive hope is not in technology. We use technology for specific purposes. But Chesterton also pushes us in a healthy direction. While techno-fetishism, as an end in itself, is never the goal, the solution is not to dismiss the smartphone. Chesterton forbids us from becoming blind to the breathtaking power of our phones, which can be traced back to the glory of God. We approach a text message beep not with a *tsk* of irritation over the intrusion but as a new prompt for healthy wonder.

GRATEFUL TO TEARS

My life will be governed by one of two perspectives: God-centered awe, in a world soaked with his glory and governed by his sovereign presence, or technological atheism, buffered from God, with faith in the right techniques and controls to govern the reality of a disen-chanted and mechanistically driven world.[10] This is the decision we face each time we pick up our phones.

Toward the end of my research for this book, I asked John Piper how he uses technology in fulfilling the purpose and calling of his life, and he was quick to gush over all the ways his apps and Bible software have fed his soul over the years. At the end, he looked down at his laptop, his iPad, and his iPhone, all sitting on the table, and he said, "I could almost come to tears over how precious they are to me." Yes, they are glowing tools made mostly by men and women who are not submitted to God, he reiterated, and they are tools that open up his life to a thousand convenient temptations, but used with care and discipline, the digital tools are, he said, "a treasure chest of the glories of God."[11]

I deeply desire his discipline—to use my phone as a means to genuinely encounter God, to gratefully tap its full eternal value. But for many of us, who lack this maturity, technology tends to feed our

9. Richard Stivers, *Shades of Loneliness: Pathologies of a Technological Society* (Lanham, MD: Rowman & Littlefield, 2004), 121.

10. See Tony Reinke, "The Rise of the Modern Control Freak," tonyreinke.com (March 16, 2016).

11. John Piper, interview with the author via Skype, published as "How Do You Use Your iPhone and iPad in Christian Growth?" Desiring God, desiringGod.org (April 1, 2016).

vanity and kill our wonder. At worst, our phones are handheld wands of power that promise to protect our sinful isolation, showcase our self-aggrandizement, prop up our digital towers of self-praise, feed our materialism, lure us to so-called "anonymous" vices, and offer an "escape" from our creaturehood. We cannot marvel at technology by abusing it. True wonder requires humility. Wonder is the special joy of God reserved for those who have become childlike and humbled under the awe of a divine Father. In humility, we become "wonderers," freed from secular disenchantment, from commercially driven promises that materialism cannot deliver, and from temporal entrapments in order to more clearly behold God's glory in and through our technology.

When we use our smartphones rightly, their shining screens radiate with the treasure of God's glory in Christ, and in that glory-glow, we get a sneak peek into a greater age to come.

FORTHCOMING

Christ reigns sovereignly over all technology, but all technology has not yet been subjected to his moral will.[12] When that day comes, God will unveil his holy city before our eyes. This city will be free of all sin, and it will be full of innovations that seem almost unimaginable now. The apostle John gives us a preview:

> The wall was built of jasper, while the city was pure gold, like clear glass. The foundations of the wall of the city were adorned with every kind of jewel. The first was jasper, the second sapphire, the third agate, the fourth emerald, the fifth onyx, the sixth carnelian, the seventh chrysolite, the eighth beryl, the ninth topaz, the tenth chrysoprase, the eleventh jacinth, the twelfth amethyst. And the twelve gates were twelve pearls, each of the gates made of a single pearl, and the street of the city was pure gold, like transparent glass. (Rev. 21:18–21)

This splendid new city will be wrested from the best raw materials

---

12. Heb. 2:8.

of the earth, its precious jewels and materials will shine the brilliant light of the new creation—Jesus Christ glorified—and it will dazzle![13] The sun in our sky is a mere placeholder for a glory to come that will be sevenfold times brighter, and all of this new creation will be true art, serving to make the resplendence of the Son even more brilliant and stunning.

How superb this vision must have been in John's day, in Israel, where landscapes were made of dust, homes were made of stones, streets were made of mud, and midnight darkness was unabated.

Perhaps our imaginations have a jump-start on the images? Centuries of technological innovation make the vision of John seem more realistic to us. We can begin to imagine a world with no true night, because we have electricity to shine lights 24/7, even for outdoor sporting events. Our urban skyscrapers are gilded with glass, reflecting glorious sunsets in a golden skyline. We can begin to understand a solid city, radiant with glory and able to embrace light.

But our world is still only a dim reflection. Our strongest walls and our best roads are steel, concrete, and black asphalt, not clear gold. Glass cannot carry the weight of our cities, and until it does, glory cannot pass through our world like the new world we read about. For now, we are illuminated constantly by the flickering rectangles of our devices, but not continuously by the glory of Christ's physical presence.

The point of Scripture is that the wicked city Babylon and all of its godless machinery will be uprooted and cast away to make room for God's city, the New Jerusalem, shining with sights and technologies now unimaginable and exceeding all human ingenuity and expectation.[14]

---

13. Isa. 60:19; Rev. 21:11, 23; 22:5.

14. Isa. 60:17–22. Does the vision in Revelation suggest a generally heightened technological advance in the new creation? Yes and no. I *do* think John saw innovations he could not fully put into words, but I am not suggesting the jewels embedded in the walls are really touch screens, LED lights, or anything resembling the technological advances we are familiar with today simply projected into his vision. We must refrain both from imagining a techless eternity that is completely unrelated to human innovation and from thinking that what makes heaven *heaven* is its techno-culture. The Lamb's omnipresent glory will be the centerpiece of eternity, and all future technology will serve him. Yet I also believe redemptive history suggests that we should expect

Our greatest need in the digital age is to behold the glory of the unseen Christ in the faint blue glow of our pixelated Bibles, by faith.[15] But in the new creation, in God's finished city, we will enjoy the blazing splendor of Christ, by sight. This moment will mark the pinnacle of our lives, when we are transfigured into perfect image bearers of God.[16] In this beatific vision, our souls will be ravished and joy will spill over from our hearts forever in a night-free eternity. The new creation will fulfill Jesus's longing and prayer that we would dwell with him, not merely for a splendid moment,[17] but permanently, in the light of his unfading glory.[18]

BACK TO THE FUTURE

So, no, it is not odd to end our journey with the candlestick telephone. Our technologies pass in and out of style in a flash, and the smartphone gadget we hold onto dearly today will soon be discredited in light of newer and better innovations. In the end, all of our current devices will be tossed aside when God reveals his master plan for where technology was leading all along.

I'm not saying that technology is worthless; I am certainly not saying every technology will live on in eternity; and I sincerely hope that technologies and social platforms that were en vogue when I wrote this book will pass into obsolescence and make room for new ones—better ones—at our fingertips. Our future selves will look back at our present selves and laugh about our glowing rectangles, wired earbuds, tangled charging wires, and limited batteries. The

---

continuity between the technological advances in human history and the technology that will appear in eternity, including such things as air travel. If human dominion and labor exist in the heavenly world, and I believe they do (Matt. 25:14–30), then I must also believe eternity will continue to provide us a stage where we will unveil endless technological advances in a way far superior to anything we know on earth, but not dislocated from the technological trajectory we are witnessing now. Grace will purify and perfect our technological intentions, though we must leave much ambiguity in our predictions.

15. 2 Cor. 3:12–4:6.
16. 1 John 3:2.
17. Matt. 17:1–8; Mark 9:2–8; Luke 9:28–36.
18. John 17:24. The most beautiful sustained meditation on this passage is found in John Owen's *Meditations and Discourses on the Glory of Christ*, in *The Works of John Owen*, ed. William H. Goold (Edinburgh: Banner of Truth, 1965), 1:273–461.

clunky devices we now tote as breaking innovations will be nearly as unrecognizable to our grandchildren as cassette tapes are to my children. We are fooling about with smartphones as finite creatures who can live only in space and time. We may pinch our faces at Chesterton's awe over the candlestick phone, but the fact remains: the smartphone is the techno-wonder of our day. It is celebrated as the most influential gadget in human history. And yet it is passing away. It is becoming obsolete. And we are called to live smartphone smart as we, in Christ, move toward a resplendent city full of glory and innovation that will blur our smartphones into a foggy memory.

ACKNOWLEDGMENTS

Wise living in the smartphone age requires untangling an enormously complex knot of issues, and for that I needed help. The urgency of the task demanded many Christian minds to come together and humbly work toward thoughtful solutions. This book happened because a team of friends pitched in, and all of them deserve special mention. (Although as time exposes the errors and shortsightedness of this book, those are mine to own.)

Special thanks to the theologians who agreed to answer questions about technology in interviews with me, and none more than John Piper, who has answered hundreds of my questions in the daily *Ask Pastor John* podcast. Pastor John, your willingness to tackle a dozen questions on technology for the podcast (and ultimately for readers of this project) was greatly appreciated. I love your example, appreciate your friendship, and thank God for the way you model the works of exegesis, theology, and ethics with due earnestness.

Special thanks to several other theologians, historians, philosophers, ethicists, pastors, and artists who agreed to interviews with me on various topics of technology: Francis Chan, Matt Chandler, Sinclair Ferguson, Craig Gay, Douglas Groothuis, Bruce Hindmarsh, Tim Keller, Trip Lee, Peter Leithart, Richard Lints, Oliver O'Donovan, Ray Ortlund, David Powlison, Alastair Roberts, Kevin Vanhoozer, and David Wells. (I have not quoted all of these men from those key interviews, but transcripts are collected at tonyreinke.com.)

Smartphones and social media will never stop changing, and that means wisdom for Christian living in the digital age will always be the product of ongoing dialogue between Christian leaders, parents, friends, and local churches. I hope this book helps spark many new conversations. The friends who dialogued with me over this book include Joe and Sylvie Osburn, who invested an amazing amount of thought and critical interaction, as did John Dyer, Janice Evans, Tracy Fruehauf, Gloria Furman, Jasmine Holmes, Brad Littlejohn, David Murray, Kim Cash Tate, and Liz Wann. Thank you all.

Thank you to four key editors: Alastair Roberts, Paul Maxwell, Jon Vickery, and Bryan DeWire, a researcher's best friend, who managed a digital bushel's worth of relevant articles on technology.

A general thank you to the entire Crossway team for their continued excellence at every level of publishing, and a special thanks to my advisor and friend Justin Taylor for making this project happen. And thank you to my wise editor and strategic redactor Greg Bailey, who enriched every page of this book. It is an honor to work with you and to benefit from your editorial excellence.

The cover design calls for huge thanks to visionary designer Josh Dennis of Crossway, who imagined iPhone Guy; to Don Clark at Invisible Creature Inc., who designed him; and to Curtis Clark, an expert guitar builder, who brought him into being with his woodworking skills. The cover image will never fully represent his scale (iPhone Guy stands six feet tall and weighs 250 pounds!). Josh, Don, Curtis—I am grateful to God for your expertise, innovation, and artistic genius.

Thanks to our three precious kids—Jon, Christabel, and Bunyan. When I write books in quiet, I often imagine I'm speaking to you, or, more accurately, writing to your future selves. So I want to thank you (your present selves) for supporting and loving me while I wrote this, and to say hello to and thank you (your future selves) for reading it. I love you.

Above all, I thank my Savior and my God for the grace I have been given. And next in line, I thank you, Karalee, my wife. Perhaps I don't understand how dedications work, because this is my third book dedicated to you—and it's hard to think of anyone more deserving. You believe in me, support my labors in endless ways, and edit every sentence with great skill to make sense of what I'm trying to say. I can only write books with you beside me.

GENERAL INDEX

SCRIPTURE INDEX